Scattering Analysis of Periodic Structures Using Finite-Difference Time-Domain Method

Synthesis Lectures on Computational Electromagnetics

Editor
Constantine A. Balanis, ***Arizona State University***

Synthesis Lectures on Computational Electromagnetics will publish 50- to 100-page publications on topics that include advanced and state-of-the-art methods for modeling complex and practical electromagnetic boundary value problems. Each lecture develops, in a unified manner, the method based on Maxwell's equations along with the boundary conditions and other auxiliary relations, extends underlying concepts needed for sequential material, and progresses to more advanced techniques and modeling. Computer software, when appropriate and available, is included for computation, visualization and design. The authors selected to write the lectures are leading experts on the subject that have extensive background in the theory, numerical techniques, modeling, computations and software development.
The series is designed to:

- Develop computational methods to solve complex and practical electromagnetic boundary-value problems of the 21st century.

- Meet the demands of a new era in information delivery for engineers, scientists, technologists and engineering managers in the fields of wireless communication, radiation, propagation, communication, navigation, radar, RF systems, remote sensing, and biotechnology who require a better understanding and application of the analytical, numerical and computational methods for electromagnetics.

Scattering Analysis of Periodic Structures Using Finite-Difference Time-Domain Method
Khaled ElMahgoub, Fan Yang, and Atef Elsherbeni
2012

Introduction to the Finite-Difference Time-Domain (FDTD) Method for Electromagnetics
Stephen D. Gedney
2011

Analysis and Design of Substrate Integrated Waveguide Using Efficient 2D Hybrid Method
Xuan Hui Wu and Ahmed A. Kishk
2010

An Introduction to the Locally-Corrected Nyström Method
Andrew F. Peterson and Malcolm M. Bibby
2009

Transient Signals on Transmission Lines: An Introduction to Non-Ideal Effects and Signal Integrity Issues in Electrical Systems
Andrew F. Peterson and Gregory D. Durgin
2008

Reduction of a Ship's Magnetic Field Signatures
John J. Holmes
2008

Integral Equation Methods for Electromagnetic and Elastic Waves
Weng Cho Chew, Mei Song Tong, and Bin Hu
2008

Modern EMC Analysis Techniques Volume II: Models and Applications
Nikolaos V. Kantartzis and Theodoros D. Tsiboukis
2008

Modern EMC Analysis Techniques Volume I: Time-Domain Computational Schemes
Nikolaos V. Kantartzis and Theodoros D. Tsiboukis
2008

Particle Swarm Optimization: A Physics-Based Approach
Said M. Mikki and Ahmed A. Kishk
2008

Three-Dimensional Integration and Modeling: A Revolution in RF and Wireless Packaging
Jong-Hoon Lee and Manos M. Tentzeris
2007

Electromagnetic Scattering Using the Iterative Multiregion Technique
Mohamed H. Al Sharkawy, Veysel Demir, and Atef Z. Elsherbeni
2007

Electromagnetics and Antenna Optimization Using Taguchi's Method
Wei-Chung Weng, Fan Yang, and Atef Elsherbeni
2007

Fundamentals of Electromagnetics 1: Internal Behavior of Lumped Elements
David Voltmer
2007

Fundamentals of Electromagnetics 2: Quasistatics and Waves
David Voltmer
2007

Modeling a Ship's Ferromagnetic Signatures
John J. Holmes
2007

Mellin-Transform Method for Integral Evaluation: Introduction and Applications to Electromagnetics
George Fikioris
2007

Perfectly Matched Layer (PML) for Computational Electromagnetics
Jean-Pierre Bérenger
2007

Adaptive Mesh Refinement for Time-Domain Numerical Electromagnetics
Costas D. Sarris
2006

Frequency Domain Hybrid Finite Element Methods for Electromagnetics
John L. Volakis, Kubilay Sertel, and Brian C. Usner
2006

Exploitation of A Ship's Magnetic Field Signatures
John J. Holmes
2006

Support Vector Machines for Antenna Array Processing and Electromagnetics
Manel Martínez-Ramón and Christos Christodoulou
2006

The Transmission-Line Modeling (TLM) Method in Electromagnetics
Christos Christopoulos
2006

Computational Electronics
Dragica Vasileska and Stephen M. Goodnick
2006

Higher Order FDTD Schemes for Waveguide and Antenna Structures
Nikolaos V. Kantartzis and Theodoros D. Tsiboukis
2006

Introduction to the Finite Element Method in Electromagnetics
Anastasis C. Polycarpou
2006

MRTD(Multi Resolution Time Domain) Method in Electromagnetics
Nathan Bushyager and Manos M. Tentzeris
2006

Mapped Vector Basis Functions for Electromagnetic Integral Equations
Andrew F. Peterson
2006

Scattering Analysis of Periodic Structures Using Finite-Difference Time-Domain Method
Khaled ElMahgoub, Fan Yang, and Atef Elsherbeni

ISBN: 978-3-031-00585-5 paperback
ISBN: 978-3-031-01713-1 ebook

DOI 10.1007/978-3-031-01713-1

A Publication in the Springer series
SYNTHESIS LECTURES ON COMPUTATIONAL ELECTROMAGNETICS

Lecture #28
Series Editor: Constantine A. Balanis, *Arizona State University*
Series ISSN
Synthesis Lectures on Computational Electromagnetics
Print 1932-1252 Electronic 1932-1716

Scattering Analysis of Periodic Structures Using Finite-Difference Time-Domain Method

Khaled ElMahgoub, Fan Yang, and Atef Elsherbeni
University of Mississippi

SYNTHESIS LECTURES ON COMPUTATIONAL ELECTROMAGNETICS #28

ABSTRACT

Periodic structures are of great importance in electromagnetics due to their wide range of applications such as frequency selective surfaces (FSS), electromagnetic band gap (EBG) structures, periodic absorbers, meta-materials, and many others. The aim of this book is to develop efficient computational algorithms to analyze the scattering properties of various electromagnetic periodic structures using the finite-difference time-domain periodic boundary condition (FDTD/PBC) method. A new FDTD/PBC-based algorithm is introduced to analyze general skewed grid periodic structures while another algorithm is developed to analyze dispersive periodic structures. Moreover, the proposed algorithms are successfully integrated with the generalized scattering matrix (GSM) technique, identified as the hybrid FDTD-GSM algorithm, to efficiently analyze multilayer periodic structures. All the developed algorithms are easy to implement and are efficient in both computational time and memory usage. These algorithms are validated through several numerical test cases. The computational methods presented in this book will help scientists and engineers to investigate and design novel periodic structures and to explore other research frontiers in electromagnetics.

KEYWORDS

finite difference time domain (FDTD), periodic structures, periodic boundary conditions (PBC), generalized scattering matrix (GSM), frequency selective surfaces (FSS), multi-layer structures, auxiliary differential equation (ADE), dispersive material, general skewed grid

Khaled ElMahgoub: To my family and friends

Fan Yang: To my family and colleagues

Atef Elsherbeni: To my family

Contents

Preface

This book is intended to help students, researchers, and engineers who are using electromagnetics tools to investigate and design novel periodic structures and to explore related research frontiers in electromagnetics. Various electromagnetic periodic structures, such as dispersive materials, multi-layered structures, and arbitrary skewed grids, are studied using the finite-difference time-domain with periodic boundary condition (FDTD/PBC) method. The book starts with a description of the FDTD approach and the constant horizontal wavenumber PBC technique. The main advantages and limitations of the approach are discussed in Chapter 2. The FDTD updating equations are derived and numerical results are provided to verify the proposed approach. In Chapter 3, the constant horizontal wavenumber approach is extended to analyze periodic structures with an arbitrary skewed grid. The new approach is described and the FDTD updating equations are derived for both cases in which the skewed shift is coincident and non-coincident with the FDTD grid. In Chapter 4, a new dispersive periodic boundary condition (DPBC) for the FDTD technique is developed. The algorithm utilizes the auxiliary differential equation (ADE) technique with two-term Debye relaxation equation to simulate the general dispersive property in the medium. In addition, the constant horizontal wavenumber technique is modified accordingly to model the dispersive material on the periodic boundaries. In Chapter 5, a complete analysis of multi-layer periodic structures using the hybrid FDTD/GSM method is illustrated. Based on the FDTD simulation results on each layer, the generalized scattering matrix (GSM) cascading technique is used to analyze different kinds of multilayered periodic structures.

The algorithms developed in this book are implemented using MATLAB. These algorithms lead to comprehensive software tools that are capable of analyzing efficiently and accurately general electromagnetic periodic structures. These software tools can be used in many design applications involving different configurations and types of periodic structures with ordinary and dispersive-type material.

Khaled ElMahgoub, Fan Yang, and Atef Elsherbeni
May 2012

Acknowledgments

The authors would like to thank Allen Glisson, Chair and Professor of Electrical Engineering, University of Mississippi, Kai-Fong Lee, Professor of Electrical Engineering, University of Mississippi, William Staton, Professor of Mathematics, University of Mississippi, Veysel Demir, Assistant Professor of Electrical Engineering, Northern Illinois University, and Ji Chen, Associate Professor of Electrical and Computer Engineering, University of Houston, for their help in reviewing the material for this book and their productive collaboration in this research. Finally, the authors would like to thank Morgan & Claypool for their patience as we completed this work.

Khaled ElMahgoub, Fan Yang, and Atef Elsherbeni
May 2012

CHAPTER 1

Introduction

1.1 BACKGROUND

The finite-difference time-domain (FDTD) method has gained great popularity as an effective tool for solving Maxwell's equations. The FDTD method is based on a simple formulation that does not require complex asymptotic or Green's functions. Although it is a time-domain simulation, it provides a wideband frequency-domain response using time-domain to frequency-domain transformation. It can easily handle composite structures consisting of different types of materials. In addition, it can be easily implemented using parallel computational algorithms. These features of FDTD have made it one of the most attractive techniques in computational electromagnetics for many applications. FDTD has been used to solve numerous types of problems such as scattering, microwave circuits, waveguides, antennas, propagation, non-linear and other special materials, and many other applications [1].

Periodic electromagnetic structures are of great importance due to their applications in frequency selective surfaces (FSS), electromagnetic band gap (EBG) structures, corrugated surfaces, phased antenna arrays, periodic absorbers, negative index materials, etc. Many versions of the FDTD algorithms have been developed to analyze such structures and to make use of the periodicity of these structures. Periodic boundary conditions (PBC) have been implemented in many forms such that only one unit cell needs to be analyzed instead of the entire structure. These techniques are divided into two main categories: field transformation methods and direct field methods [2]. Field transformation methods introduce auxiliary fields to eliminate the need for time-advanced data; the transformed field equations are then discretized and solved using FDTD techniques. The split-field method [3] and multi-spatial grid method [4] are useful approaches in this category. There are two main limitations with these methods. First, the transformed equations have additional terms that require special handling such as splitting the field or using a multi-grid algorithm to implement the FDTD, which increases the complexity of the algorithm. Second, as the angle of incidence increases from normal incidence ($\theta = 0°$) to grazing incidence ($\theta = 90°$), the stability factor needs to be reduced, so the FDTD time step decreases significantly [2]. As a result, a larger number of time steps are needed for oblique incidence to generate stable results, which increases the computational time for such cases.

As for the direct field category, these methods work directly with Maxwell's equations, and there is no need for any field transformation. An example of these methods is the sine-cosine method [5], in which the structure is excited simultaneously with sine and cosine waveforms. The PBC for oblique incidence can be implemented using this method. The stability criterion for this technique is the same as the conventional FDTD (angle-independent), which provides stable analysis for incidence near grazing. However, it is a single frequency method and loses an important property of FDTD, the wide-band capability.

In [6, 7, 8] a simple and efficient FDTD/PBC algorithm was introduced that belongs to the direct field category and yet has a wideband capability. In this approach, the FDTD simulation is performed by setting a constant horizontal wavenumber instead of a specific angle of incidence. The idea of using a constant wavenumber in FDTD was originated from guided wave analysis and eigenvalue problems in [9], and it was extended to the plane wave scattering problems in [10, 11, 12]. The approach offers many advantages, such as implementation simplicity, stability condition and numerical errors similar to the conventional FDTD, computational efficiency near the grazing incident angles, and the wide-band capability. Due to the advantages offered by the constant horizontal wavenumber PBC, it is used in this book as a basis to develop new algorithms that solve challenges in the simulation of diversified periodic structures, such as skewed grid periodic structures, dispersive periodic structures, and multi-layer periodic structures.

1.2 CONTRIBUTIONS

This book starts with a description of the FDTD constant horizontal wavenumber approach. The main advantages and limitations of the approach are discussed. The FDTD updating equations are derived and numerical results are provided to demonstrate the validity of the approach.

It's worthwhile to point out that most previous FDTD PBCs were developed to analyze axial grid periodic structure. However, there are numerous applications where the grid of the periodic structures is a general skewed grid. In Chapter 3, the constant horizontal wavenumber approach is extended to analyze periodic structures with an arbitrary skewed grid. The new approach is described and the FDTD updating equations are derived for both cases in which the skewed shift is coincident and non-coincident with the FDTD grid. Numerical results are presented to prove the validity of the new approach.

Furthermore, it is noticed that most previous PBCs for the FDTD technique were developed to analyze periodic structures where dispersive materials are not located on the boundaries of the unit cell. However, there are some applications where periodic structures with dispersive media on the boundaries of the unit cell must be used. In Chapter 4, a new dispersive periodic boundary condition (DPBC) for the FDTD technique is developed to solve the above challenge. The algorithm utilizes the auxiliary differential equation (ADE) technique with two-term Debye relaxation equation to simulate the general dispersive property in the medium. In addition, the constant horizontal wavenumber approach is modified accordingly to implement the periodic boundary conditions. The validity of this algorithm is verified through several numerical examples.

In today's applications, many periodic structures are often built up of layers, each layer being either a diffraction grating, periodic in one or two directions, or a homogenous dielectric slab that acts as a separator or support. In Chapter 5, a complete analysis of a multi-layer periodic structure using the hybrid FDTD/GSM method is illustrated. Based on the FDTD simulation results on each layer, the generalized scattering matrix (GSM) cascading technique is used to analyze different kinds of multi-layer periodic structures. In addition, a complete Floquet harmonic analysis of periodic structure is presented, where propagation and evanescent behaviors of Floquet harmonics are studied. Moreover, guidelines for harmonics selection are provided. Different cases of multi-layer periodic structures are analyzed and numerical results are presented to prove the validity and efficiency of the new proposed algorithms.

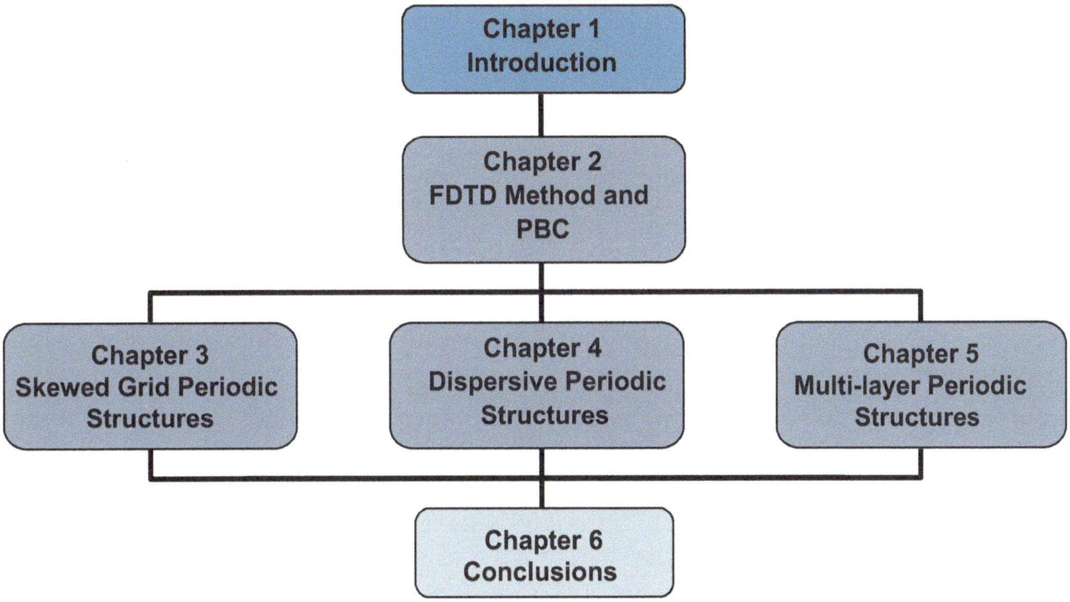

Figure 1.1: Book chapters layout.

CHAPTER 2

FDTD Method and Periodic Boundary Conditions

2.1 BASIC EQUATIONS OF THE FDTD METHOD

The FDTD method belongs to the general class of grid-based differential time-domain numerical modeling methods. The time-domain Maxwell's equations can be stated as follows:

$$\nabla \times \vec{H} = \frac{\partial \vec{D}}{\partial t} + \vec{J}, \tag{2.1a}$$

$$\nabla \times \vec{E} = -\frac{\partial \vec{B}}{\partial t} - \vec{M}, \tag{2.1b}$$

$$\nabla \cdot \vec{D} = \rho_e, \tag{2.1c}$$

$$\nabla \cdot \vec{B} = \rho_m, \tag{2.1d}$$

where $\vec{E}$ is the electric field intensity vector in V/m, $\vec{D}$ is the electric displacement vector in C/m^2, $\vec{H}$ is the magnetic field intensity vector in A/m, $\vec{B}$ is the magnetic flux density vector in Weber/m^2, $\vec{J}$ electric current density vector in A/m^2, $\vec{M}$ is the magnetic current density vector in V/m^2, ρ_e is the electric charge density in C/m^3, and ρ_m is the magnetic charge density in Weber/m^3. For linear, isotropic, and non-dispersive materials, the electric displacement vector and the magnetic flux density vector can be written as

$$\vec{D} = \varepsilon \vec{E}, \tag{2.2a}$$

$$\vec{B} = \mu \vec{H}, \tag{2.2b}$$

where ε is permittivity and μ is permeability of the material. The electric current density $\vec{J}$ is the sum of the conduction current density $\vec{J_C} = \sigma^e \vec{E}$ and the impressed current density $\vec{J_I}$ as $\vec{J} = \vec{J_C} + \vec{J_I}$. Similarly, for the magnetic current density $\vec{M} = \vec{M_C} + \vec{M_I}$, where $\vec{M_C} = \sigma^m \vec{H}$. Here σ^e is the electric conductivity of the material in S/m and σ^m is the magnetic conductivity of the material in Ω/m. Using the two curl Equations (2.1) and the Equation (2.2), Maxwell's curl equations can be rewritten as:

$$\nabla \times \vec{H} = \varepsilon \frac{\partial \vec{E}}{\partial t} + \sigma^e \vec{E} + \vec{J_I}, \tag{2.3a}$$

$$\nabla \times \vec{E} = -\mu \frac{\partial \vec{H}}{\partial t} - \sigma^m \vec{H} - \vec{M_I}. \tag{2.3b}$$

Equation (2.3) consists of two vector equations and each vector equation can be decomposed into three scalar equations in the three-dimensional space. Therefore, Maxwell's curl equations can be represented with six scalar equations in a Cartesian coordinate system (x, y, z) as follows:

$$\frac{\partial E_x}{\partial t} = \frac{1}{\varepsilon_x}\left[\frac{\partial H_z}{\partial y} - \frac{\partial H_y}{\partial z} - \sigma_x^e E_x - J_{ix}\right], \tag{2.4a}$$

$$\frac{\partial E_y}{\partial t} = \frac{1}{\varepsilon_y}\left[\frac{\partial H_x}{\partial z} - \frac{\partial H_z}{\partial x} - \sigma_y^e E_y - J_{iy}\right], \tag{2.4b}$$

$$\frac{\partial E_z}{\partial t} = \frac{1}{\varepsilon_z}\left[\frac{\partial H_y}{\partial x} - \frac{\partial H_x}{\partial y} - \sigma_z^e E_z - J_{iz}\right], \tag{2.4c}$$

$$\frac{\partial H_x}{\partial t} = \frac{1}{\mu_x}\left[\frac{\partial E_y}{\partial z} - \frac{\partial E_z}{\partial y} - \sigma_x^m H_x - M_{ix}\right], \tag{2.4d}$$

$$\frac{\partial H_y}{\partial t} = \frac{1}{\mu_y}\left[\frac{\partial E_z}{\partial x} - \frac{\partial E_x}{\partial z} - \sigma_y^m H_y - M_{iy}\right], \tag{2.4e}$$

$$\frac{\partial H_z}{\partial t} = \frac{1}{\mu_z}\left[\frac{\partial E_x}{\partial y} - \frac{\partial E_y}{\partial x} - \sigma_z^m H_z - M_{iz}\right]. \tag{2.4f}$$

The material parameters ε_x, ε_y, and ε_z are associated with electric field components E_x, E_y, and E_z, respectively, through Equation (2.2a). Similarly, the material parameters μ_x, μ_y, and μ_z are associated with magnetic field components H_x, H_y, and H_z, respectively, through Equation (2.2b) as pointed out in [1].

The first step in the FDTD algorithm is approximating the time and space derivatives appearing in Maxwell's equations by finite-differences. The central finite-difference scheme is used here as an approximation of the space and time derivatives of both the electric and magnetic fields. For example the derivative of a function $f(x)$ at a point x_0 using central finite-difference can be written as

$$f'(x_0) \approx \frac{f(x_0 + \Delta x) - f(x_0 - \Delta x)}{2\Delta x}, \tag{2.5}$$

where Δx is the sampling period.

Secondly, the electric and the magnetic field components are assigned to certain positions in each cell. In 1966, Yee was the first to set up the commonly used arrangement of these field components to solve both the electric and magnetic Maxwell's curl equations in an iterative time sequence [13].

For the Yee cell shown in Fig. 2.1, the three components of electric and magnetic fields are placed in certain positions in the cell, such that the electric field vectors form loops around the magnetic field vectors, which simulates Faraday's law and magnetic field vectors form loops around the electric field vectors, which simulates Ampere's law. The electric field vectors are assigned to the center of the edges of the cells, while the magnetic field vectors are assigned to the center of the faces of the cells. The calculations of the electric and magnetic fields are not only offset in position but also in time. The electric field components are calculated at a certain time instant $(n+1)\,\Delta t$, while the magnetic field components are calculated at the time instant $(n+0.5)\,\Delta t$.

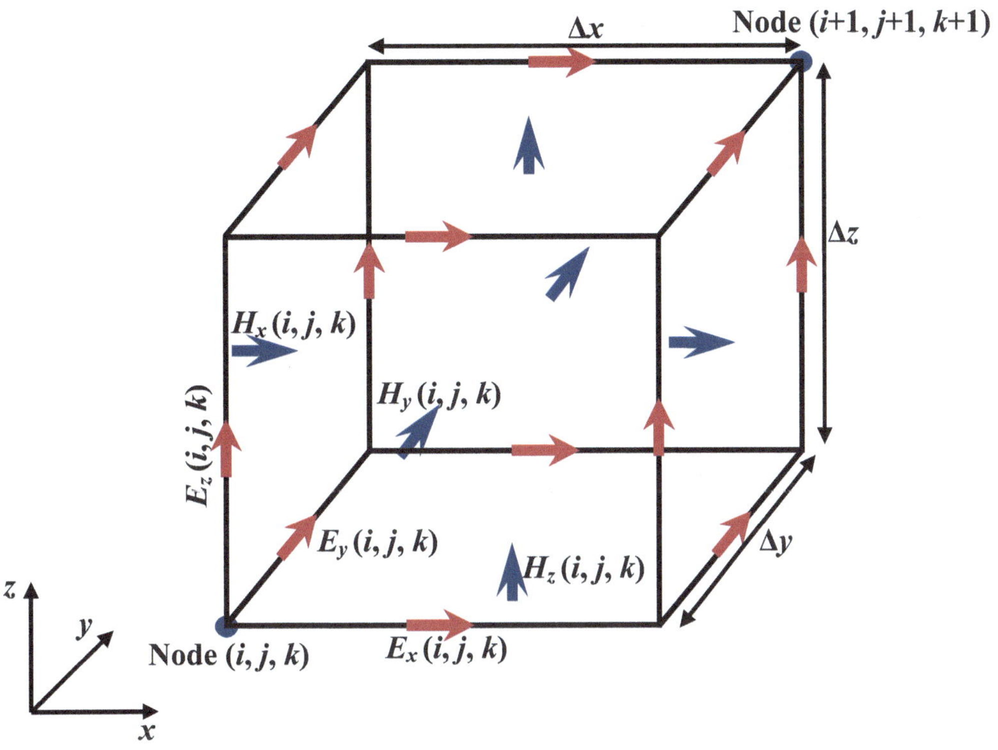

Figure 2.1: Arrangement of field components at node (i, j, k) base on Yee's cell indexing scheme.

Equations (2.4) and (2.5) are used to construct six scalar FDTD updating equations for the six components of electromagnetic fields by the introduction of respective coefficient terms as follows [1]:

For the E_x component:

$$\begin{aligned} E_x^{n+1}(i,j,k) =& C_{exe}(i,j,k) \times E_x^n(i,j,k) \\ &+ C_{exhz}(i,j,k) \times \left[H_z^{n+\frac{1}{2}}(i,j,k) - H_z^{n+\frac{1}{2}}(i,j-1,k) \right] \\ &+ C_{exhy}(i,j,k) \times \left[H_y^{n+\frac{1}{2}}(i,j,k) - H_y^{n+\frac{1}{2}}(i,j,k-1) \right] \\ &+ C_{exj}(i,j,k) \times J_{ix}^{n+\frac{1}{2}}(i,j,k), \end{aligned} \tag{2.6}$$

where

$$C_{exe}(i,j,k) = \frac{2\varepsilon_x(i,j,k) - \Delta t\sigma_x^e(i,j,k)}{2\varepsilon_x(i,j,k) + \Delta t\sigma_x^e(i,j,k)}, C_{exhz}(i,j,k) = \frac{2\Delta t}{(2\varepsilon_x(i,j,k) + \Delta t\sigma_x^e(i,j,k))\Delta y},$$
$$C_{exhy}(i,j,k) = \frac{-2\Delta t}{(2\varepsilon_x(i,j,k) + \Delta t\sigma_x^e(i,j,k))\Delta z}, C_{exj}(i,j,k) = \frac{-2\Delta t}{2\varepsilon_x(i,j,k) + \Delta t\sigma_x^e(i,j,k)}.$$

For the E_y component:

$$\begin{aligned}E_y^{n+1}(i,j,k) =& C_{eye}(i,j,k) \times E_y^n(i,j,k) \\ &+ C_{eyhx}(i,j,k) \times \left[H_x^{n+\frac{1}{2}}(i,j,k) - H_x^{n+\frac{1}{2}}(i,j,k-1) \right] \\ &+ C_{eyhz}(i,j,k) \times \left[H_z^{n+\frac{1}{2}}(i,j,k) - H_z^{n+\frac{1}{2}}(i-1,j,k) \right] \\ &+ C_{eyj}(i,j,k) \times J_{iy}^{n+\frac{1}{2}}(i,j,k),\end{aligned} \tag{2.7}$$

where

$$C_{eye}(i,j,k) = \frac{2\varepsilon_y(i,j,k) - \Delta t\sigma_y^e(i,j,k)}{2\varepsilon_y(i,j,k) + \Delta t\sigma_y^e(i,j,k)}, C_{eyhx}(i,j,k) = \frac{2\Delta t}{(2\varepsilon_y(i,j,k) + \Delta t\sigma_y^e(i,j,k))\Delta z},$$
$$C_{eyhz}(i,j,k) = \frac{-2\Delta t}{(2\varepsilon_y(i,j,k) + \Delta t\sigma_y^e(i,j,k))\Delta x}, C_{eyj}(i,j,k) = \frac{-2\Delta t}{2\varepsilon_y(i,j,k) + \Delta t\sigma_y^e(i,j,k)}.$$

For the E_z component:

$$\begin{aligned}E_z^{n+1}(i,j,k) =& C_{eze}(i,j,k) \times E_z^n(i,j,k) \\ &+ C_{ezhy}(i,j,k) \times \left[H_y^{n+\frac{1}{2}}(i,j,k) - H_y^{n+\frac{1}{2}}(i-1,j,k) \right] \\ &+ C_{ezhx}(i,j,k) \times \left[H_x^{n+\frac{1}{2}}(i,j,k) - H_x^{n+\frac{1}{2}}(i,j-1,k) \right] \\ &+ C_{ezj}(i,j,k) \times J_{iz}^{n+\frac{1}{2}}(i,j,k),\end{aligned} \tag{2.8}$$

where

$$C_{eze}(i,j,k) = \frac{2\varepsilon_z(i,j,k) - \Delta t\sigma_z^e(i,j,k)}{2\varepsilon_z(i,j,k) + \Delta t\sigma_z^e(i,j,k)}, C_{ezhy}(i,j,k) = \frac{2\Delta t}{(2\varepsilon_z(i,j,k) + \Delta t\sigma_z^e(i,j,k))\Delta x},$$
$$C_{ezhx}(i,j,k) = \frac{-2\Delta t}{(2\varepsilon_z(i,j,k) + \Delta t\sigma_z^e(i,j,k))\Delta y}, C_{ezj}(i,j,k) = \frac{-2\Delta t}{2\varepsilon_z(i,j,k) + \Delta t\sigma_z^e(i,j,k)}.$$

For the H_x component:

$$\begin{aligned} H_x^{n+\frac{1}{2}}(i,j,k) =& C_{hxh}(i,j,k) \times H_x^{n-\frac{1}{2}}(i,j,k) \\ &+ C_{hxey}(i,j,k) \times \left[E_y^n(i,j,k+1) - E_y^n(i,j,k) \right] \\ &+ C_{hxez}(i,j,k) \times \left[E_z^n(i,j+1,k) - E_z^n(i,j,k) \right] \\ &+ C_{hxm}(i,j,k) \times M_{ix}^n(i,j,k), \end{aligned} \tag{2.9}$$

where

$$C_{hxh}(i,j,k) = \frac{2\mu_x(i,j,k) - \Delta t \sigma_x^m(i,j,k)}{2\mu_x(i,j,k) + \Delta t \sigma_x^m(i,j,k)}, \quad C_{hxey}(i,j,k) = \frac{2\Delta t}{(2\mu_x(i,j,k) + \Delta t \sigma_x^m(i,j,k))\Delta z},$$

$$C_{hxez}(i,j,k) = \frac{-2\Delta t}{(2\mu_x(i,j,k) + \Delta t \sigma_x^m(i,j,k))\Delta y}, \quad C_{hxm}(i,j,k) = \frac{-2\Delta t}{2\mu_x(i,j,k) + \Delta t \sigma_x^m(i,j,k)}.$$

For the H_y component:

$$\begin{aligned} H_y^{n+\frac{1}{2}}(i,j,k) =& C_{hyh}(i,j,k) \times H_y^{n-\frac{1}{2}}(i,j,k) \\ &+ C_{hyez}(i,j,k) \times \left[E_z^n(i+1,j,k) - E_z^n(i,j,k) \right] \\ &+ C_{hyex}(i,j,k) \times \left[E_x^n(i,j,k+1) - E_x^n(i,j,k) \right] \\ &+ C_{hym}(i,j,k) \times M_{iy}^n(i,j,k), \end{aligned} \tag{2.10}$$

where

$$C_{hyh}(i,j,k) = \frac{2\mu_y(i,j,k) - \Delta t \sigma_y^m(i,j,k)}{2\mu_y(i,j,k) + \Delta t \sigma_y^m(i,j,k)}, \quad C_{hyez}(i,j,k) = \frac{2\Delta t}{(2\mu_y(i,j,k) + \Delta t \sigma_y^m(i,j,k))\Delta x},$$

$$C_{hyex}(i,j,k) = \frac{-2\Delta t}{(2\mu_y(i,j,k) + \Delta t \sigma_y^m(i,j,k))\Delta z}, \quad C_{hym}(i,j,k) = \frac{-2\Delta t}{2\mu_y(i,j,k) + \Delta t \sigma_y^m(i,j,k)}.$$

For the H_z component:

$$\begin{aligned} H_z^{n+\frac{1}{2}}(i,j,k) =& C_{hzh}(i,j,k) \times H_z^{n-\frac{1}{2}}(i,j,k) \\ &+ C_{hzex}(i,j,k) \times \left[E_x^n(i,j+1,k) - E_x^n(i,j,k) \right] \\ &+ C_{hzey}(i,j,k) \times \left[E_y^n(i+1,j,k) - E_y^n(i,j,k) \right] \\ &+ C_{hzm}(i,j,k) \times M_{iz}^n(i,j,k), \end{aligned} \tag{2.11}$$

where

$$C_{hzh}(i,j,k) = \frac{2\mu_z(i,j,k) - \Delta t \sigma_z^m(i,j,k)}{2\mu_z(i,j,k) + \Delta t \sigma_z^m(i,j,k)}, \quad C_{hzex}(i,j,k) = \frac{2\Delta t}{(2\mu_z(i,j,k) + \Delta t \sigma_z^m(i,j,k))\Delta y},$$

$$C_{hzey}(i,j,k) = \frac{-2\Delta t}{(2\mu_z(i,j,k) + \Delta t \sigma_z^m(i,j,k))\Delta x}, \quad C_{hzm}(i,j,k) = \frac{-2\Delta t}{2\mu_z(i,j,k) + \Delta t \sigma_z^m(i,j,k)}.$$

After deriving the six FDTD updating Equations (2.6)–(2.11), a time-marching algorithm can be constructed, as shown in Fig. 2.2.

The first step in this algorithm is setting up the problem space, including objects, material types, sources, etc., and defining any other parameters such as the excitation waveforms that will be used during the FDTD computation. The problem space usually has a finite size and specific boundary conditions can be enforced on the boundaries of the problem space. Therefore, the field components on the boundaries of the problem are treated according to the type of the boundary conditions during the iteration. After the fields are updated and boundary conditions are enforced, the current values of the desired field components are captured and stored as output data, and this data can be used for real time processing and/or post-processing in order to calculate other desired parameters. The FDTD iterations can be continued until certain stopping criteria are achieved.

2.2 PERIODIC BOUNDARY CONDITIONS

The speed and the storage space of a simulation depend mainly on the size of the computational domain. For a free space scattering problem, the computational domain needs to be extended to infinity, which means an infinite number of cells in the computational domain is needed. The solution of this problem is to truncate the domain by a set of artificial boundaries at a certain distance from the objects. Various boundary conditions were developed to solve this problem, such as perfect electric conductor (PEC) boundaries, which can be used to simulate cavity structures, and absorbing boundary conditions (ABC), which can be used to simulate open boundary problems. Different methods have been used to simulate an absorbing boundary condition in FDTD calculations. The most common ones are Mur's approach [14], Liao's approach [15], perfectly matched layer (PML) [16], and convolutional perfect matched layer (CPML) [17].

Periodic boundary conditions (PBCs) were developed to analyze periodic structures in FDTD simulations. The main idea is to make use of the periodic nature of the structure such that only one unit cell needs to be analyzed instead of the entire structure. The difference between a periodic boundary condition and a normal absorbing boundary condition is that electric field components outside the boundary are known for PBC due to the periodicity. Consider the 1-D periodic problem shown in Fig. 2.3, the fields in the unit cell (i+N+1) can be readily determined by the fields in unit cell (i+N), so do the fields in unit cell (i+N+2), etc.

From the above figure, it is obvious that E_{zn+1} component can be updated using the knowledge of E_{z0}. According to the Floquet theory, the boundary electric field of a periodic structure with periodicity P_x along the x-direction can be written in the frequency-domain as

$$E(x=0, y, z, \omega) = E(x=P_x, y, z, \omega) \times e^{jk_x P_x}, \tag{2.12}$$

where k_x is the propagation constant in the x direction. Assuming the angle of the incident plane wave is θ, the horizontal wavenumber k_x is given by

$$k_x = k_0 \sin\theta \tag{2.13}$$

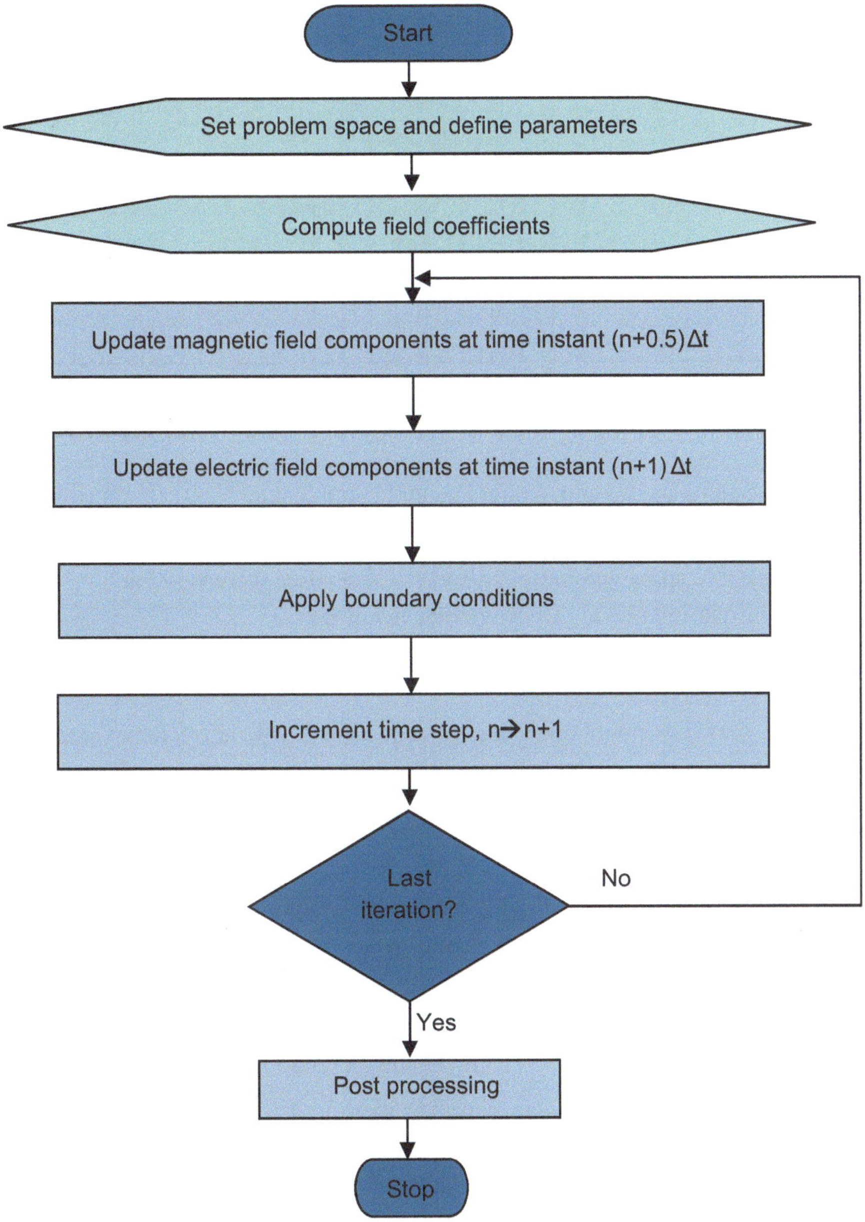

Figure 2.2: The flowchart of the conventional FDTD code [1].

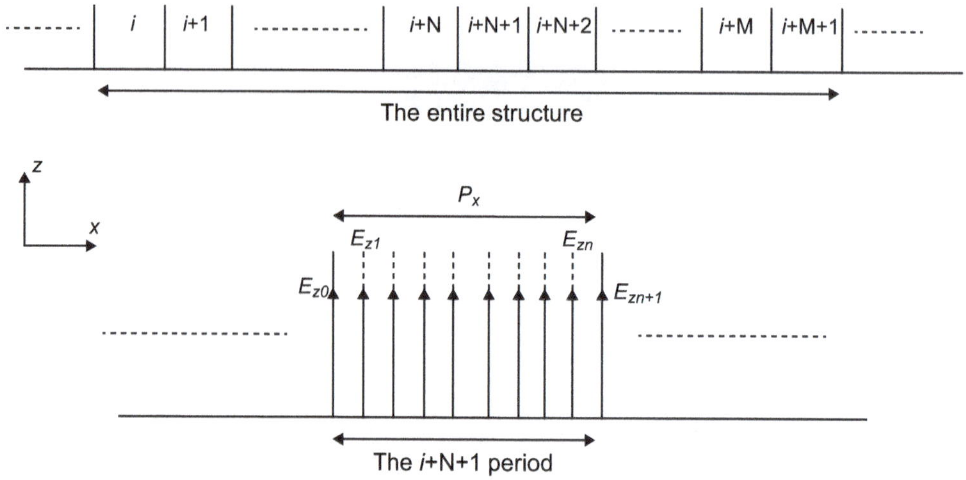

Figure 2.3: 1-D periodic structure with periodicity P_x in x-direction.

where $k_0 = \omega/c$ is the free space wavenumber, and c is free space wave velocity.

Using (2.13) Equation (2.12) can be written as follows:

$$E(x = 0, y, z, \omega) = E(x = P_x, y, z, \omega) \times e^{jP_x \frac{\omega}{c} \sin\theta}. \tag{2.14}$$

One should notice that the exponential term in Equation (2.14) is frequency dependent. Using frequency-domain to time-domain transformation (2.14) can be written as follows:

$$E(x = 0, y, z, t) = E(x = P_x, y, z, t + \frac{P_x}{c} \sin\theta). \tag{2.15}$$

From (2.15) a time-advanced electric field components is needed to update Maxwell's equations at time t, which require a special handling. An explanation for different techniques handling this challenge is provided in the next section.

2.3 CONSTANT HORIZONTAL WAVENUMBER APPROACH

To understand the constant horizontal wavenumber method, the case of an infinite dielectric slab shown in Fig. 2.4 (a) is used as an example. The slab is illuminated with a TM^z (transverse magnetic) plane wave. The thickness of the slab in the z-direction is h = 0.2 m, and its relative permittivity is ε_r = 4; the reflection coefficient of the infinite slab is shown in Fig. 2.4 (b). From the figure, it should be noticed that the reflection coefficient plotted in the k_x-frequency plane provides a complete description of the scattering properties of the dielectric slab for all angles of incidence. In addition, Fig. 2.4 illustrates different FDTD methods: the solid line represents the split-field method, which simulates oblique incidence with a fixed incident angle and a band of frequencies; the small star represents the sine-cosine method, which simulates the oblique incidence at fixed incident angle and a fixed frequency; the dotted vertical line represents normal incidence; and the dashed line represents the constant horizontal wavenumber method, which simulates the oblique incidence at a fixed propagation constant k_x, which indicates different angles of incidence at different frequencies. From Fig. 2.4 it could also be noticed that for a certain k_x value the simulation is only valid from a certain minimum frequency on the light line.

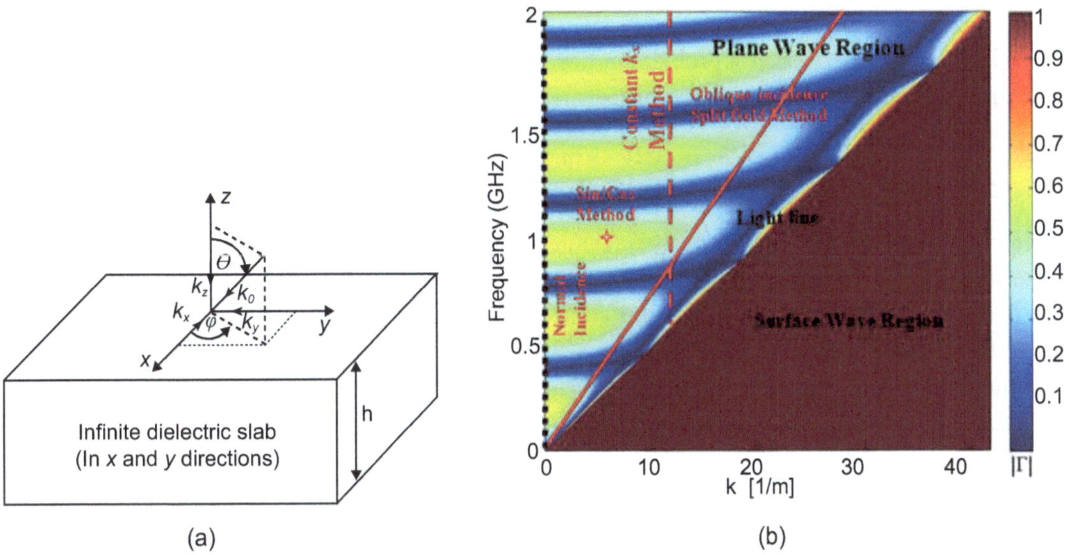

Figure 2.4: (a) k_0 direction and slab geometry, (b) Analytical reflection coefficient of infinite dielectric slab presented in the k_x-frequency plane.

The constant horizontal wavenumber approach is to fix the value of the horizontal wavenumber k_x in the FDTD simulation instead of the angle θ, where k_x is determined by both frequency and angle of incidence. Thus, the term $e^{jk_x P_x}$ is considered as a complex constant in (2.12). Using direct frequency to time-domain transformation, the field in the time-domain can be represented as follows:

$$E(x = 0, y, z, t) = E(x = P_x, y, z, t) \times e^{jk_x P_x}. \tag{2.16}$$

It should also be pointed out that both the $\boldsymbol{E}$ and $\boldsymbol{H}$ fields have complex values in the FDTD computational domain because of the PBC in (2.16) [6].

Therefore, by fixing k_x (varying angle with frequency), the need for the knowledge of time-advanced electric field components to update Maxwell's equations is eliminated. An important issue related to the constant wavenumber method is the plane wave excitation procedure. If the traditional total-field/scattered-field (TF/SF) formulation described in [18] is applied, a problem arises regarding the incident angle. For example, the tangential electric field component of a TM^z

(transverse magnetic) incident wave depends on the incident angle. To overcome this problem, the TF/SF technique is modified. In the case of TM^z excitation, only the tangential magnetic incident field component is imposed on the excitation plane $z = z_0$. This one-field excitation allows the plane wave to propagate in both directions $z > z_0$ and $z < z_0$ (z_0 is the excitation plane position). Thus, the entire computational domain becomes the total field region, and there is no scattered field region. The scattered field can be calculated using the difference between the total and the incident field. Similarly, for the TE^z (transverse electric) case, only the tangential electric incident field component is imposed.

In addition, there exists a problem of horizontal resonance, where fields do not decay to zero over time. To avoid this problem, the proper frequency range for the excitation waveform must be chosen as follows [6]:

$$f_C = \frac{k_x c}{2\pi} + \frac{BW}{2}. \tag{2.17}$$

where f_C is the center frequency of the Gaussian pulse and BW is the bandwidth of the Gaussian pulse. In this approach, the conventional Yee scheme shown in Fig. 2.1 can be used to update the $\boldsymbol{E}$ and $\boldsymbol{H}$ fields, which offers several advantages, such as implementation simplicity and the same stability condition and numerical errors similar to the conventional FDTD. In addition, the computational efficiency for incident angles near grazing and the wideband capability are achieved as well [6]. This makes the constant horizontal wavenumber approach a good choice for the analysis of periodic structures.

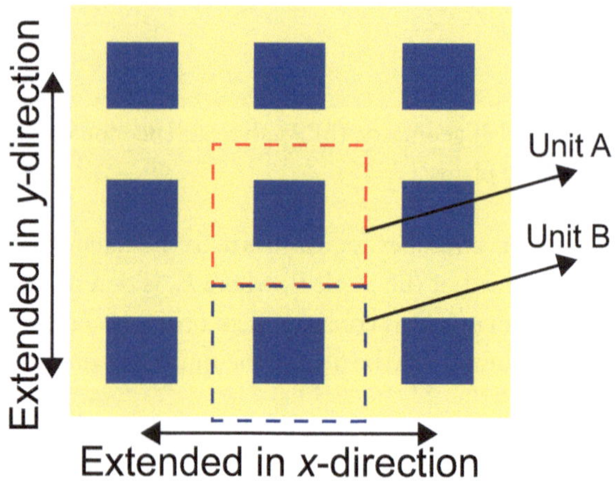

Figure 2.5: Periodic structure geometry (square patch FSS).

With the proposed periodic boundary condition (PBC), the reflection and transmission properties of the periodic structure shown in Fig. 2.5 can be calculated using FDTD by simulating the unit cell A only. The magnetic field components are updated using the conventional FDTD updating

Equations (2.9)–(2.11), while the non-boundary components of the electric field will be updated using the conventional FDTD updating Equations (2.6)–(2.8). The components on the boundaries will be updated using the PBC equations based on the constant horizontal wavenumber approach. Thus, the updating equations for the boundary electric field components are organized as follows:

1) Updating E_x at $y = 0$ and $y = P_y$.

2) Updating E_y at $x = 0$ and $x = P_x$.

3) Updating E_z at $y = 0$, $y = P_y$, $x = 0$, and $x = P_x$, except for the corners.

4) Updating E_z at the corners.

1) To update the E_x on the boundary $y = 0$, the magnetic field components H_z outside the unit A are needed. However, due to the periodicity in the y-direction, one can use the magnetic field components H_z of interest inside unit A to update these electric fields such that

$$\begin{aligned} E_x^{n+1}(i,1,k) =& C_{exe}(i,1,k) \times E_x^n(i,1,k) + C_{exhz}(i,1,k) \times [H_z^{n+1/2}(i,1,k) - H_z^{n+1/2}(i,0,k)] \\ &+ C_{exhy}(i,1,k) \times [H_y^{n+1/2}(i,1,k) - H_y^{n+1/2}(i,1,k-1)], \end{aligned} \tag{2.18}$$

where the coefficients are stated as in (2.6), and $H_z^{n+1/2}(i,0,k) = H_z^{n+1/2}(i,n_y,k) \times e^{jk_yP_y}$ due to the periodicity in the y-direction as shown in Fig. 2.6. The term $e^{jk_yP_y}$ is used to compensate the phase shift due to the oblique incidence.

Then the updating equation for E_x on the boundary $y = 0$ can be written as

$$\begin{aligned} E_x^{n+1}(i,1,k) =& C_{exe}(i,1,k) \times E_x^n(i,1,k) \\ &+ C_{exhz}(i,1,k) \times [H_z^{n+1/2}(i,1,k) - H_z^{n+1/2}(i,n_y,k) \times e^{jk_yP_y}] \\ &+ C_{exhy}(i,1,k) \times [H_y^{n+1/2}(i,1,k) - H_y^{n+1/2}(i,1,k-1)]. \end{aligned} \tag{2.19}$$

As for E_x on the boundary $y = P_y$, the updating equation can be written as

$$E_x^{n+1}(i,n_y+1,k) = E_x^{n+1}(i,1,k) \times e^{-jk_yP_y}. \tag{2.20}$$

2) Due to periodicity in x-direction as shown in Fig. 2.7, the updating equation for the E_y component on the boundary $x = 0$ can be written as

$$\begin{aligned} E_y^{n+1}(1,j,k) =& C_{eye}(1,j,k) \times E_y^n(1,j,k) \\ &+ C_{eyhx}(1,j,k) \times [H_x^{n+1/2}(1,j,k) - H_x^{n+1/2}(1,j,k-1)] \\ &+ C_{eyhz}(1,j,k) \times [H_z^{n+1/2}(1,j,k) - H_z^{n+1/2}(n_x,j,k) \times e^{jk_xP_x}]. \end{aligned} \tag{2.21}$$

As for E_y on the boundary $x = P_x$, the updating equation can be written as

$$E_y^{n+1}(n_x+1,j,k) = E_y^{n+1}(1,j,k) \times e^{-jk_xP_x}. \tag{2.22}$$

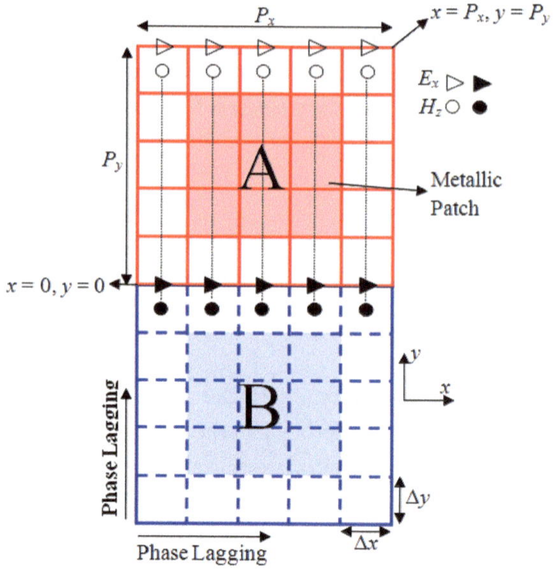

Figure 2.6: FDTD grid for unit A and the adjacent unit B.

3) A similar procedure is used for updating the E_z components, but corner components are updated separately due to the presence of the periodicity in both x- and y-directions. For E_z on the boundaries $x = 0$ and $x = P_x$, the updating equation can be written for $j \neq 1$ and $j \neq n_y + 1$ (avoiding the corners) as

$$\begin{aligned} E_z^{n+1}(1, j, k) = & C_{eze}(1, j, k) \times E_z^n(1, j, k) \\ & + C_{ezhy}(1, j, k) \times [H_y^{n+\frac{1}{2}}(1, j, k) - H_y^{n+\frac{1}{2}}(n_x, j, k) \times e^{jk_x P_x}] \\ & + C_{ezhx}(1, j, k) \times [H_x^{n+\frac{1}{2}}(1, j, k) - H_x^{n+\frac{1}{2}}(1, j-1, k)], \end{aligned} \tag{2.23}$$

$$E_z^{n+1}(n_x + 1, j, k) = E_z^{n+1}(1, j, k) \times e^{-jk_x P_x}. \tag{2.24}$$

The updating equation for the E_z components on the boundaries $y = 0$, and $y = P_y$ can be written for $i \neq 1$ and $i \neq n_x + 1$ (avoiding the corners) as

$$\begin{aligned} E_z^{n+1}(i, 1, k) = & C_{eze}(i, 1, k) \times E_z^n(i, 1, k) \\ & + C_{ezhy}(i, 1, k) \times [H_y^{n+\frac{1}{2}}(i, 1, k) - H_y^{n+\frac{1}{2}}(i-1, 1, k)] \\ & + C_{ezhx}(i, 1, k) \times [H_x^{n+\frac{1}{2}}(i, 1, k) - H_x^{n+\frac{1}{2}}(i, n_y, k) \times e^{jk_y P_y}], \end{aligned} \tag{2.25}$$

$$E_z^{n+1}(i, n_y + 1, k) = E_z^{n+1}(i, 1, k) \times e^{-jk_y P_y}. \tag{2.26}$$

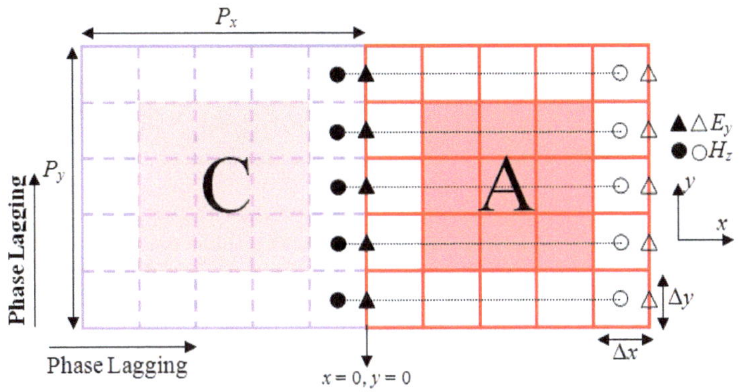

Figure 2.7: FDTD grid for unit A and the adjacent unit C, E_y components.

4) The E_z components at the corners are updated as follows:
At $x = 0$ and $y = 0$,

$$\begin{aligned} E_z^{n+1}(1,1,k) =& C_{eze}(1,1,k) \times E_z^n(1,1,k) \\ &+ C_{ezhy}(1,1,k) \times [H_y^{n+\frac{1}{2}}(1,1,k) - H_y^{n+\frac{1}{2}}(n_x,1,k) \times e^{jk_x P_x}] \\ &+ C_{ezhx}(1,1,k) \times [H_x^{n+\frac{1}{2}}(1,1,k) - H_x^{n+\frac{1}{2}}(1,n_y,k) \times e^{jk_y P_y}]. \end{aligned} \tag{2.27}$$

At $x = P_x$ and $y = 0$,

$$E_z^{n+1}(n_x+1,1,k) = E_z^{n+1}(1,1,k) \times e^{-jk_x P_x}. \tag{2.28}$$

At $x = 0$ and $y = P_y$,

$$E_z^{n+1}(1,n_y+1,k) = E_z^{n+1}(1,1,k) \times e^{-jk_y P_y}. \tag{2.29}$$

At $x = P_x$ and $y = P_y$,

$$E_z^{n+1}(n_x+1,n_y+1,k) = E_z^{n+1}(1,1,k) \times e^{-jk_y P_y} \times e^{-jk_x P_x}. \tag{2.30}$$

Equations (2.18)–(2.30) describe the necessary discretization equation used in the constant horizontal wavenumber method. All these equations are derived for an obliquely incident plane wave. For a normally incident plane wave all the phase compensation terms should be set to one. After deriving the updating equations, a time marching algorithm can be constructed as shown in Fig. 2.8. The main difference between this algorithm and the conventional FDTD algorithm shown in Fig. 2.2, is the updating of the boundary electric field components.

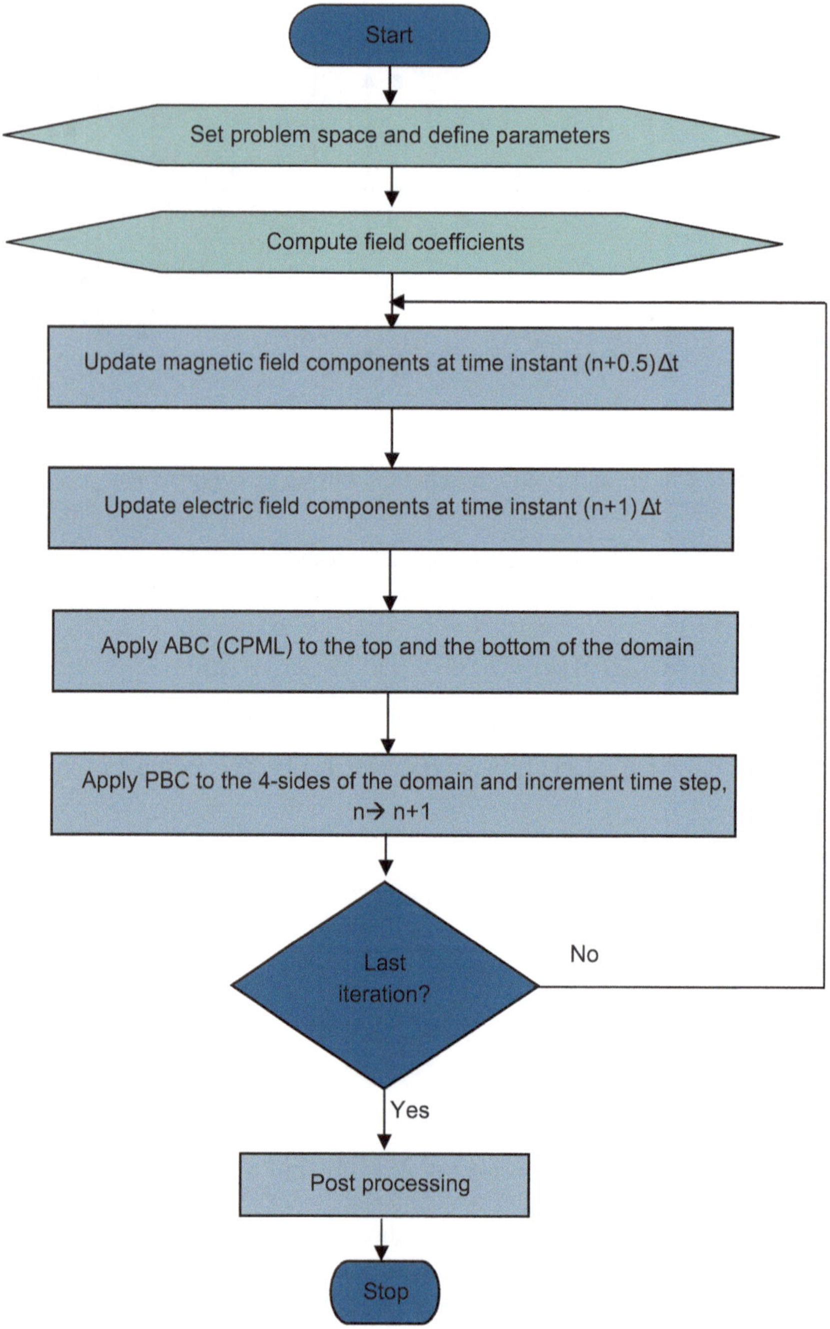

Figure 2.8: The flowchart of the FDTD/PBC code.

2.4 NUMERICAL RESULTS

In this section, numerical results generated using the constant horizontal wavenumber method are presented. The FDTD code was developed using MATLAB [19]. All the test cases were executed using the same computer (Intel Core 2 CPU 6700 2.66 GHz with 2 GB RAM). These results demonstrate the validity of the approach for determining reflection and transmission properties of periodic structures. The first example is an infinite dielectric slab excited by TM^z and TE^z plane waves. The second example is a dipole FSS, and the last example is a Jerusalem cross (JC) FSS. The results are compared with results obtained from analytical solutions for the dielectric slab and Ansoft Designer [20] (which is based on method of moments (MoM)) solutions for the dipole and JC FSS. The numerical results are shown in two different representations. The first representation plots results of reflection coefficient magnitude versus frequency with certain horizontal wavenumber values. The second representation plots the results of the reflection coefficient magnitude versus frequency for a certain angle of incidence, which requires multiple runs of the code to generate such results. In addition, the MATLAB code is capable of generating the phase of the reflection coefficient. Moreover, the code is capable of extracting the magnitude and phase of the transmission coefficient as well as the reflection and transmission cross-polarization coefficients.

2.4.1 AN INFINITE DIELECTRIC SLAB

Due to its homogeneity, the infinite dielectric slab can be considered as a periodic structure with any periodicity. In addition, the analytical solution can be easily generated, which makes the infinite dielectric slab an appropriate verification case. The FDTD code is first used to analyze an infinite dielectric slab with thickness h = 9.375 mm and relative permittivity ε_r = 2.56 (the reflection and transmission properties of an infinite dielectric slab can be calculated analytically). The slab is illuminated by TM^z and TE^z plane waves, respectively. The slab is excited using a cosine-modulated Gaussian pulse centered at 10 GHz with a 20 GHz bandwidth (in this book the bandwidth of modulated Gaussian pulse is defined as the frequency band where the magnitude of the frequency domain reaches 10% of its maximum). Two cases are examined where the plane wave is incident normally ($k_x = k_y = 0$ m^{-1}) in the first case and obliquely ($k_x = 104.8$ m^{-1}, $k_y = 0$ m^{-1} for a minimum frequency of 5 GHz) in the second case. The FDTD grid cell size is $\Delta x = \Delta y = \Delta z =$ 0.3125 mm and the slab is represented by 5×5 cells. In the FDTD code, 2,500 time steps and 0.9 reduction factor of the Courant–Friedrich–Levy (CFL) [21] time step as used in [1, 21].

The CPML is used as absorbing boundaries at the top and the bottom of the computational domain as shown in Fig. 2.9. In Fig. 2.10 the FDTD results are compared with the analytical solution results, where good agreement for both TM^z and TE^z cases (normal and oblique incidence) is observed. The stability of the algorithm is observed even at the angles of incidence near grazing ($\theta = 90°$).

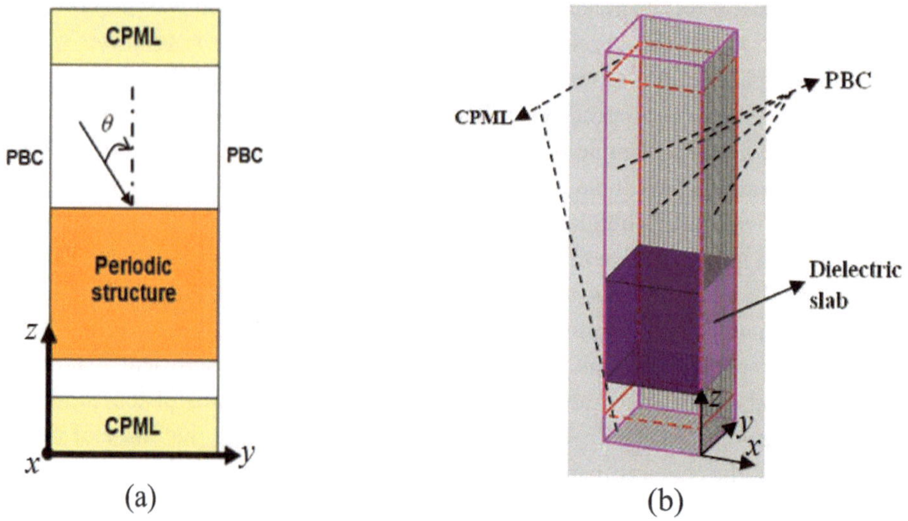

Figure 2.9: (a) The FDTD/PBC domain with different boundary conditions, (b) An infinite dielectric slab in the FDTD/PBC computational domain.

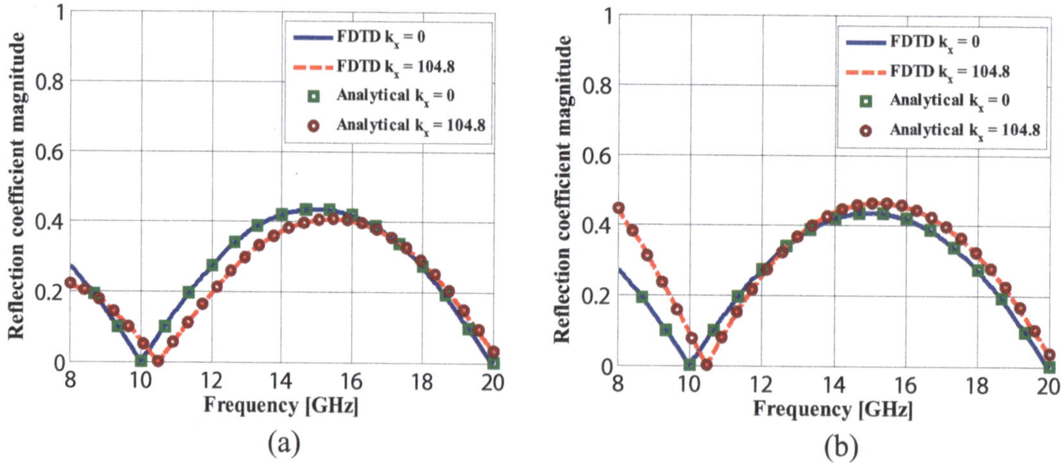

Figure 2.10: Reflection coefficient for an infinite dielectric slab, (a) TM^z case, (b) TE^z case.

2.4.2 A DIPOLE FSS

The algorithm is then used to analyze an FSS structure consisting of dipole elements. The dipole length is 12 mm and its width is 3 mm. The unit cell periodicity is 15 mm in both the x- and y-directions. The substrate has a thickness of 6 mm and relative permittivity ε_r = 2.2, as shown in Fig. 2.11 [22]. The structure is first illuminated by a normally incident plane wave (with polarization along the y-axis). Figure 2.12 provides the results for normal incidence. The structure is excited using a cosine-modulated Gaussian pulse centered at 8 GHz with a 16 GHz bandwidth.

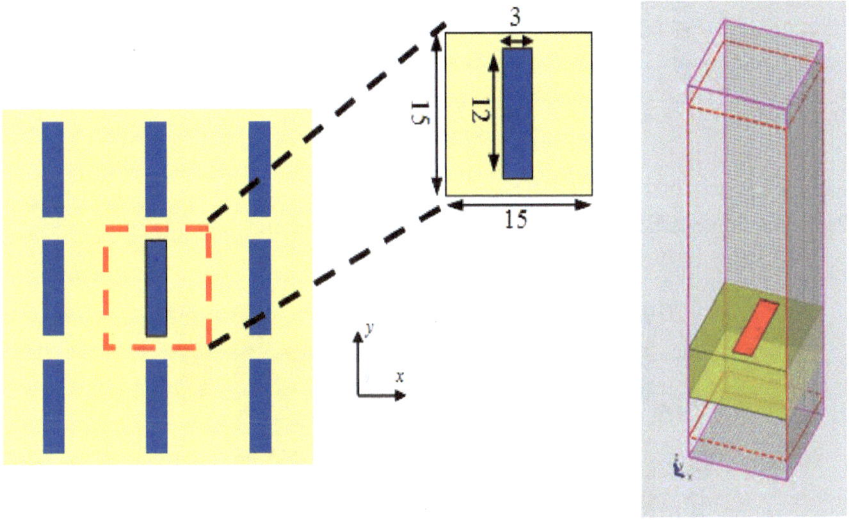

Figure 2.11: Dipole FSS geometry (all dimensions are in mm).

In the FDTD code, 2,500 time steps and a 0.9 reduction factor of the CFL time step are used. The CPML is used for absorbing boundaries at the top and the bottom of the computational domain. The FDTD grid cell size is $\Delta x = \Delta y = \Delta z = 0.5$ mm. The results are compared with results obtained using Ansoft Designer. The computational time per simulation for the FDTD code is 4.53 minutes, and the memory usage is 0.2 MB, while using Ansoft Designer the computational time per simulation is 45 minutes for 30 frequency points, and the memory usage is 21 MB.

To show the capabilities of the developed constant horizontal wavenumber FDTD MALTAB code, results for several k_x's versus frequency are generated, as shown in Fig. 2.13. From the figure the reflection and transmission regions can be clearly identified. It should be noticed that near the light line (θ= 90°, $k_x = k_0$), there exist some oscillations. This is because the excitation signal is weak at that frequency region.

Figure 2.14 provides results for an oblique incidence case (k_x = 20 m^{-1} and k_y = 7.28 m^{-1} for minimum frequency of 2 GHz) where the structure is excited using a cosine-modulated Gaussian

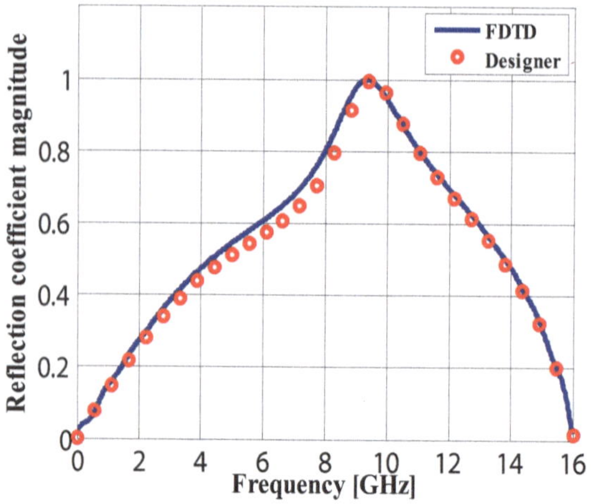

Figure 2.12: Reflection coefficient for a dipole FSS with normal incident TE^z plane wave.

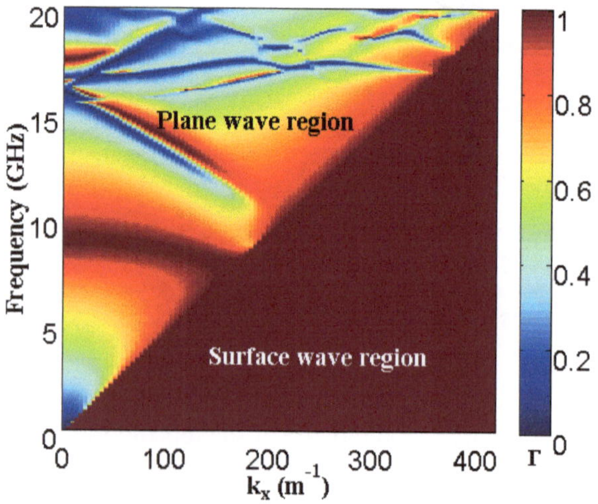

Figure 2.13: The reflection coefficient for dipole FSS with TE^z plane wave ($k_x = 0$ to 419.17 m^{-1}).

pulse centered at 9 GHz with a 14 GHz bandwidth. In the FDTD code, 2,500 time steps and a 0.9 reduction factor of CFL time step are used. The computational time per simulation for the FDTD MATLAB code is 5.023 minutes, and the memory usage is 0.9 MB, while for Ansoft Designer the computational time per simulation is 50 minutes for 30 frequency points, and the memory usage is 21 MB. The reflection coefficients for both the co-polarized and cross-polarized fields are obtained.

From Fig. 2.14 good agreement between the results generated using Ansoft Designer and results generated using the FDTD/PBC code for oblique incidence can be noticed.

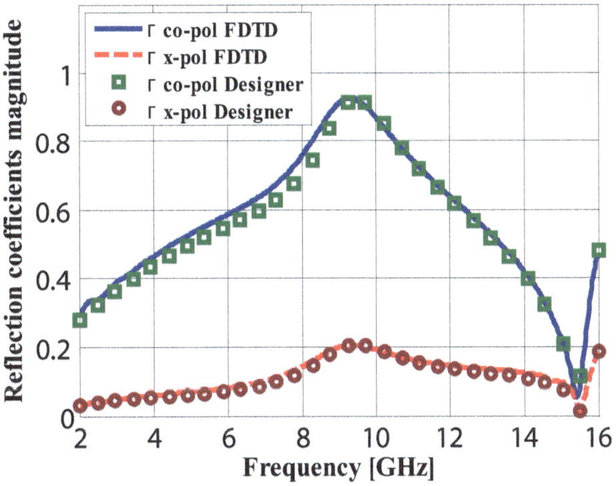

Figure 2.14: The reflection coefficient for dipole FSS with oblique incident TE^z plane wave ($k_x = 20$ m^{-1}, $ky = 7.28$ m^{-1}).

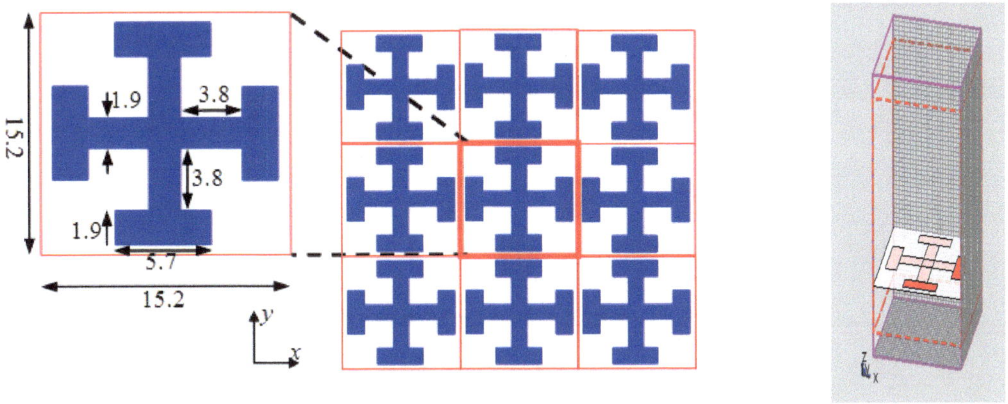

Figure 2.15: JC FSS geometry (all dimensions are in mm).

2.4.3 A JERUSALEM CROSS FSS

Next, the code based on this algorithm is used to analyze an FSS structure consisting of Jerusalem cross (JC) elements. The periodicity is 15.2 mm in both the x- and y-directions. The dimensions of the elements are shown in Fig. 2.15 [23]. The structure is illuminated by a TE^z plane wave (polarized along the y- axis). Figure 2.16 provides results for normal incidence. The structure is excited using

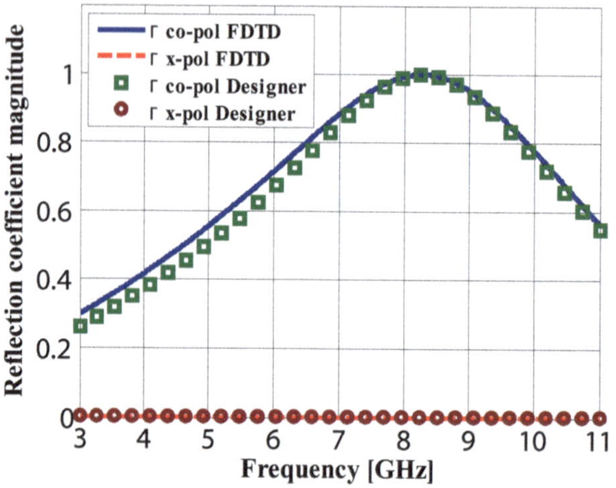

Figure 2.16: Co- and cross-polarization reflection coefficients for JC FSS with normal incident TE^z plane.

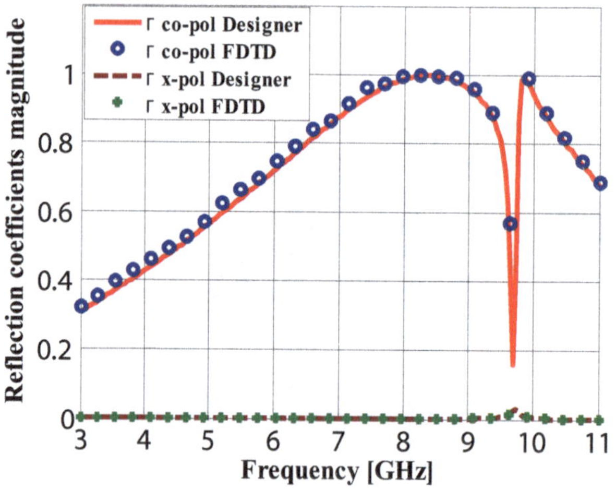

Figure 2.17: Co- and cross-polarization (Co- and x-) reflection coefficient for JC FSS with oblique incident TE^z plane wave ($\theta = 60°$), $\phi = 45°$).

a cosine-modulated Gaussian pulse centered at 7 GHz with 8 GHz bandwidth. The grid cell size is $\Delta x = \Delta y = 0.2285$ mm and $\Delta z = 0.457$ mm. In the FDTD code, 3,000 time steps and a 0.9 reduction factor of the CFL time step are used. The CPML is used for the absorbing boundaries at the top and the bottom. The results were compared with results obtained using Ansoft Designer. The computational time per simulation for the FDTD code is 5.21 minutes, and the memory usage is 0.4 MB, while for Ansoft Designer computational time per simulation is 47.5 minutes for 30 frequency points, and the memory usage is 23 MB using the same computer.

Figure 2.17 provides results for an oblique incidence ($\theta = 60°$ and $\phi = 45°$) wave exciting a JC FSS structure. To generate results for many frequency points and a specific angle of incidence, multiple runs of the code are needed, which increases the computational time.

Using 30 different k_x values (from 25.501 m^{-1} to 93.5036 m^{-1}), both co- and cross-polarization reflection coefficients were generated. The results were compared with results obtained using Ansoft Designer with good agreement as shown in Figs. 2.16 and 2.17.

2.5 SUMMARY

In this chapter, a description of the FDTD constant horizontal wavenumber approach was provided and the FDTD updating equations were derived. The approach is simple to implement and efficient in terms of both computational time and memory usage. In addition, the stability criterion is essentially angle-independent. Therefore, it is efficient in implementing incidence with angle close to grazing as well as normal incidence. It is capable of calculating the co- and cross-polarization reflection and transmission coefficients of normal and oblique incidence for both the TE^z and TM^z cases, and for different periodic structures. The numerical results show good agreement with results from the analytical solution for the dielectric slab and with those based on the MoM solutions for both dipole and JC FSS structures.

CHAPTER 3

Skewed Grid Periodic Structures

3.1 INTRODUCTION

It's worthwhile to point out that most PBCs are developed to analyze axial grid periodic structures. However, there are numerous applications where the grid of the periodic structures is a general skewed grid. For example, a triangular grid with a 60° skew angle is used in phased arrays antenna to decrease the grating lobes. Figure 3.1 shows the geometries of both axial and skewed grid structures. It is clear that the axial periodic structures are special cases of the general skewed grid structures where the skew angle $\alpha = 90°$. Although the analysis of a skewed grid periodic structure has been well developed using the MoM technique [22], it has not been fully investigated using the FDTD method. A pioneering effort presented in [24] utilizes the sine-cosine method in the analysis of periodic phased arrays with skewed grids and thus loses the wideband capability of the FDTD. Furthermore, the work presented in [24] belongs to a special case where the amount of shift in the skew direction is an integer multiple of the FDTD cell size in the same direction. This special case is referred to as "coincident" in this book.

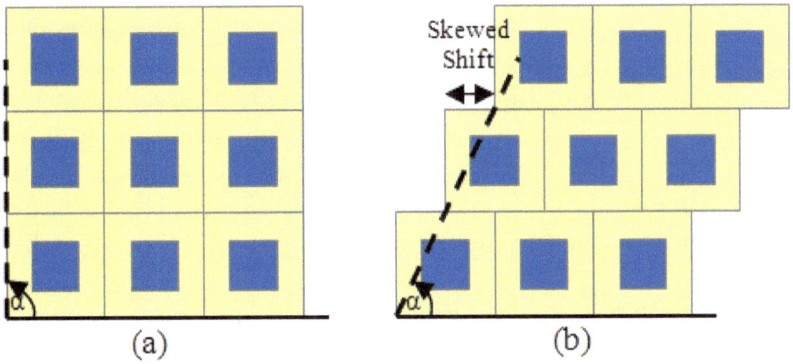

Figure 3.1: Geometries of (a) Axial, and (b) Skewed periodic structures. (From [25] © IEEE).

In this chapter, the constant horizontal wavenumber approach is extended to analyze periodic structures with skewed grids. Two types of skewed grid periodic structures are implemented. In the first category, the skew amount is coincident with the FDTD grid; and in the second category, the skew amount is non-coincident with the FDTD grid (the general skewed grid periodic structure). This chapter is organized as follows: in Section 3.2, the FDTD updating equations are derived for both the coincident and the non-coincident cases. In Section 3.3, several numerical examples proving the validity of the new approach are presented, including an infinite dielectric slab, a dipole FSS, and a Jerusalem cross FSS. Various incident angles, skew angles, and polarizations have been tested in these examples, and the numerical results show good agreement with the analytical results or other numerical results obtained from the frequency-domain methods.

3.2 CONSTANT HORIZONTAL WAVENUMBER APPROACH FOR SKEWED GRID CASE

In this section the derivation of the different electric and magnetic field updating equations are presented for two situations: in the first situation the shift is an integer number of FDTD grid cells (coincident), while in the second situation the shift is not an integer number of FDTD grid cells (non-coincident).

3.2.1 THE COINCIDENT SKEWED SHIFT

Figure 3.2 shows the FDTD grid for a coincident skewed shift periodic structure. In this specific example, the unit cell is discretized using 5x5 FDTD grid cells ($\Delta x \times \Delta y$); the unit A is the one to be simulated, while unit B and unit C are the adjacent periodic units. The structure has periodicity of P_x in the x-direction and P_y in the y-direction. S_x is the skewed shift which can be calculated as $S_x = P_y / \tan(\alpha)$, where α is the skew angle. Since the skewed shift S_x is between 0 and P_x, the skew angle is between 90° and $\tan^{-1}(P_y/P_x)$. For the periodic structures with square unit cells ($P_y = P_x$), the skew angle is between 90° and 45°. For a periodic structure with rectangular unit cells ($P_x > P_y$), it is possible to get a small skew angle.

It should be noticed from Fig. 3.2 that in this case S_x is an integer multiple of the discretization step in the x-direction (Δx). This configuration makes the shift coincident with the FDTD grid, which simplifies the calculation of the boundary electric fields. The magnetic field components are updated using the conventional FDTD updating Equations (2.9)–(2.11). As for the electric field, non-boundary components are updated using the conventional FDTD updating Equations (2.6)–(2.8).

The electric field components at the boundaries are updated using PBC equations based on the new approach. In this specific case, the skewed shift is in the x-direction. A similar procedure can be used if the skewed shift is in the y-direction [25, 26].

The updating equations for boundary electric field components are organized as follows:

1) Updating E_x at $y = 0$ and $y = P_y$

2) Updating E_y at $x = 0$ and $x = P_x$

3) Updating E_z at $y = 0$, $y = P_y$, $x = 0$, and $x = P_x$ without the corners

4) Updating E_z at the corners.

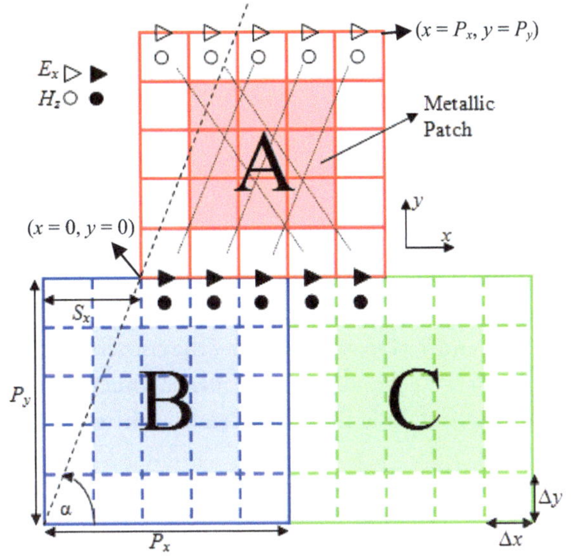

Figure 3.2: FDTD grid for skewed periodic structure coincident case, E_x components. (From [25] © IEEE).

1) To update E_x on the boundary $y = 0$, the magnetic field components H_z outside unit cell A are needed, as shown in Fig. 3.2. However, due to the periodicity and taking into account the skewed shift, one can use the magnetic field components H_z inside unit A to update these electric fields.

For $i + (S_x / \Delta x) \leq n_x$

$$H_z^{n+1/2}(i, 0, k) = H_z^{n+1/2}(i + \tfrac{S_x}{\Delta x}, n_y, k) \times e^{jk_x S_x} \times e^{jk_y P_y}, \tag{3.1}$$

while for $i + (S_x / \Delta x) > n_x$

$$H_z^{n+1/2}(i, 0, k) = H_z^{n+1/2}(i + \tfrac{S_x}{\Delta x} - n_x, n_y, k) \times e^{jk_x(S_x - P_x)} \times e^{jk_y P_y}, \tag{3.2}$$

where n_x and n_y are the total number of cells in $x-$ and y-directions, respectively. The two exponential terms are used to compensate the phase variations due to the oblique incidence. Using (3.1) and (3.2), the updating equation for the E_x components on the boundary $y = 0$ can be written as

$$\begin{aligned} E_x^{n+1}(i, 1, k) = {} & C_{exe}(i, 1, k) \times E_x^n(i, 1, k) + C_{exhz}(i, 1, k) \times [H_z^{n+1/2}(i, 1, k) - H_z^{n+1/2}(i, 0, k)] \\ & + C_{exhy}(i, 1, k) \times [H_y^{n+1/2}(i, 1, k) - H_y^{n+1/2}(i, 1, k-1)], \end{aligned} \tag{3.3}$$

where the coefficients are the same as in (2.6).

The updating equation for the E_x components on the boundary $y = P_y$ can be written for $i - (S_x/\Delta x) \leq 0$ as

$$E_x^{n+1}(i, n_y+1, k) = E_x^{n+1}(i + n_x - \tfrac{S_x}{\Delta x}, 1, k) \times e^{-jk_x(S_x - P_x)} \times e^{-jk_y P_y}, \quad (3.4)$$

while for $i - (S_x/\Delta x) > 0$

$$E_x^{n+1}(i, n_y+1, k) = E_x^{n+1}(i - \tfrac{S_x}{\Delta x}, 1, k) \times e^{-jk_x S_x} \times e^{-jk_y P_y}. \quad (3.5)$$

2) As for updating the E_y components on the boundaries $x = 0$ and $x = P_x$, (2.19) and (2.20) are used with no further modification.

3) For the E_z components on the boundaries $x = 0$ and $x = P_x$, the updating Equations (2.23) and (2.24) can be used for $j \neq 1$ and $j \neq n_y + 1$ (avoiding the corners). Updating the E_z components on the boundaries $y = 0$ and $y = P_y$ is handled in a similar manner similar to the E_x components, as shown in Fig. 3.3, which requires taking into consideration the skewed shift.

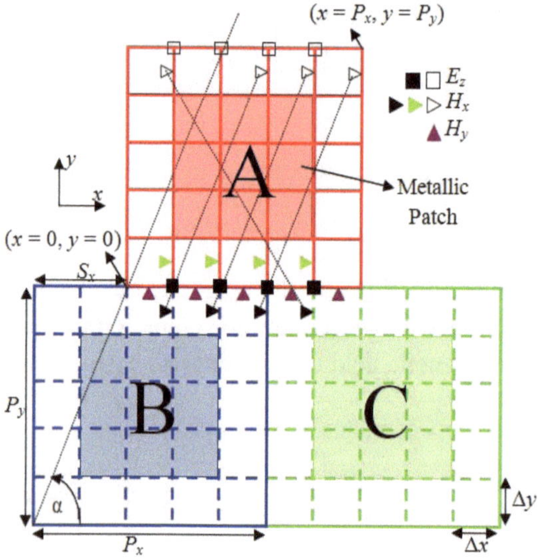

Figure 3.3: FDTD grid for coincident case, E_z components (avoiding the corners).

The updating equation for the E_z components on the boundaries $y = 0$ can be written for $i \neq 1$ and $i \neq n_x + 1$ (avoiding the corners) as

$$\begin{aligned} E_z^{n+1}(i,1,k) = {} & C_{eze}(i,1,k) \times E_z^n(i,1,k) + C_{ezhy}(i,1,k) \times [H_y^{n+1/2}(i,1,k) - H_y^{n+1/2}(i-1,1,k)] \\ & + C_{ezhx}(i,1,k) \times [H_x^{n+1/2}(i,1,k) - H_x^{n+1/2}(i,0,k)]. \end{aligned} \quad (3.6)$$

For $i + (S_x/\Delta x) \leq n_x$

$$H_x^{n+1/2}(i, 0, k) = H_x^{n+1/2}(i + \tfrac{S_x}{\Delta x}, n_y, k) \times e^{jk_x S_x} \times e^{jk_y P_y}, \quad (3.7)$$

while for $i + (S_x/\Delta x) > n_x$

$$H_x^{n+1/2}(i, 0, k) = H_x^{n+1/2}(i + \tfrac{S_x}{\Delta x} - n_x, n_y, k) \times e^{jk_x(S_x - P_x)} \times e^{jk_y P_y}. \tag{3.8}$$

4) The E_z components at the corners are updated according to Fig. 3.4 as follows:

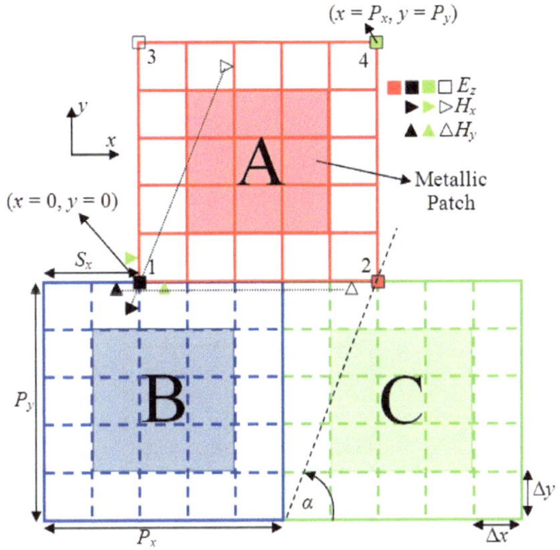

Figure 3.4: FDTD grid for coincident case, E_z corner components.

At $x = 0$ and $y = 0$

$$\begin{aligned} E_z^{n+1}(1,1,k) =& C_{eze}(1,1,k) \times E_z^n(1,1,k) \\ &+ C_{ezhy}(1,1,k) \times [H_y^{n+1/2}(1,1,k) - H_y^{n+1/2}(n_x,1,k) \times e^{jk_x P_x}] \\ &+ C_{ezhx}(1,1,k) \times [H_x^{n+1/2}(1,1,k) - H_x^{n+1/2}(1 + \frac{S_x}{\Delta x}, n_y, k) \times e^{jk_x S_x} \times e^{jk_y P_y}]. \end{aligned} \tag{3.9}$$

At $x = P_x$ and $y = 0$

$$E_z^{n+1}(n_x + 1, 1, k) = E_z^{n+1}(1, 1, k) \times e^{-jk_x P_x}. \tag{3.10}$$

At $x = 0$ and $y = P_y$

$$E_z^{n+1}(1, n_y + 1, k) = E_z^{n+1}(1 + n_x - \frac{S_x}{\Delta x}, 1, k) \times e^{-jk_x(S_x - P_x)} \times e^{-jk_y P_y}. \tag{3.11}$$

At $x = P_x$ and $y = P_y$

$$E_z^{n+1}(n_x + 1, n_y + 1, k) = E_z^{n+1}(1, n_y + 1, k) \times e^{-jk_x P_x}. \tag{3.12}$$

The above procedure provides the six FDTD updating equations for the case of a coincident skewed shift. These updating equations can be used to update electric and magnetic fields in any region in the computational domain (boundary and non-boundary).

3.2.2 THE NON-COINCIDENT SKEWED SHIFT

In this section the skewed shift is considered to have a general value, not an integer multiple of the discretization step in the x-direction (Δx) as shown in Fig. 3.5. In this case, two possible solutions can be used. The first solution is to decrease (Δx) so that the shift becomes coincident with the new discretization and one can use the formulation in the previous section, but this will increase the computational domain size and execution time. In addition, an appropriate Δx has to be chosen with every new skew angle which is not a practical solution.

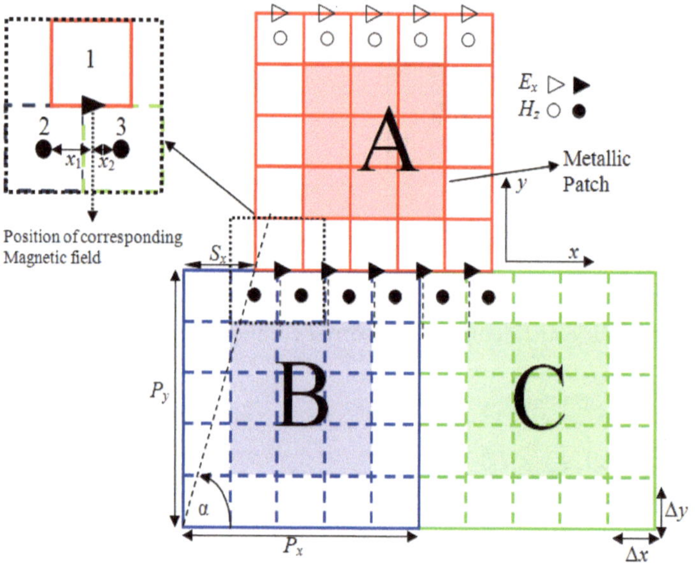

Figure 3.5: FDTD grid for skewed periodic structure non-coincident, E_x components. (From [25] © IEEE).

The second method, that will be described in this section, uses an interpolation between adjacent field components to calculate the required field component. As shown in Fig. 3.5, the shift is not an integer multiple of the discretization step in the x-direction (Δx). So the skewed shift is considered non-coincident with the FDTD grid. As a result, to update the E_x component in cell 1 (shown in the left top corner in Fig. 3.5), an interpolation between H_z in cell 2 and H_z in cell 3 is needed to get the corresponding H_z for this E_x component. The interpolation is linear interpolation based on the two distances x_1 and x_2 (x_1 is the distance between the magnetic field in cell 2 and the position of the corresponding magnetic field, and x_2 is the distance between the magnetic field in cell 3 and the position of the corresponding magnetic field). It should be noticed that the two H_z components in cells 2 and 3 are outside the unit A. However, as described in the previous section, due to periodicity and taking into account the skewed shift, one can use magnetic field components H_z inside the unit of interest to drive these two components. Then the H_z component corresponding to E_x in cell 1 can be written as

$$H_z^{n+1/2}(1,0,k)=[w_1 H_z^{n+1/2}(1+\left\lceil \tfrac{S_x}{\Delta x}\right\rceil, n_y, k)+w_2 H_z^{n+1/2}(\left\lceil \tfrac{S_x}{\Delta x}\right\rceil, n_y, k)]\times e^{jk_x S_x}\times e^{jk_y P_y}, \tag{3.13}$$

where $\lceil x \rceil$ is the ceiling function, and w_1 and w_2 are the two weighting factors calculated based on distances x_1 and x_2: $w_1 = x_1/\Delta x$, $w_2 = x_2/\Delta x$. Using (3.13) and (3.3), the $E_x(1,1,k)$ can be updated. Similarly, all other E_x components on the boundary $y = 0$ can be updated.

As for the E_x components on the boundary $y = P_y$, consider the updating equation for the first component $E_x(1, n_y+1, k)$:

$$\begin{aligned} E_x^{n+1}(1, n_y+1, k) = {} & [w_1 E_x^{n+1}(1+n_x-\left\lceil \tfrac{S_x}{\Delta x}\right\rceil, 1, k) \\ & + w_2 E_x^{n+1}(2+n_x-\left\lceil \tfrac{S_x}{\Delta x}\right\rceil, 1, k)]\times e^{-jk_x(S_x-P_x)}\times e^{-jk_y P_y}. \end{aligned} \tag{3.14}$$

Similarly, all other E_x components on the boundary $y = P_y$ can be updated. For updating the E_y components on the boundaries $x = 0$ and $x = P_x$, Equations (2.19) and (2.20) are used with no further modification. The E_z components on the boundaries $x = 0$ and $x = P_x$, the updating Equations (2.23) and (2.24) can be used for $j \neq 1$ and $j \neq n_y + 1$ (avoiding the corners).

Updating the E_z components on the boundaries $y = 0$ and $y = P_y$ (avoiding the corners) will be handled in a similar manner as the E_x components, as shown in Fig. 3.6, which requires taking into consideration the skewed shift. Note that the $H_x(i,0,k)$ is calculated from interpolation similar to the $H_z(1,0,k)$ in (3.13).

$$H_x^{n+\frac{1}{2}}(2,0,k)=[w_1 H_x^{n+\frac{1}{2}}(2+\left\lceil \frac{S_x}{\Delta x}\right\rceil, n_y, k)+w_2 H_x^{n+\frac{1}{2}}(2+\left\lceil \frac{S_x}{\Delta x}\right\rceil+1, n_y, k)]\times e^{jk_x S_x}\times e^{jk_y P_y}. \tag{3.15}$$

Using (3.15) and (3.6) the electric field $E_z(2,1,k)$ can be updated. Similarly, all other E_z components on the boundary $y = 0$ can be updated. The E_z component on the boundary $y = P_y$ can be updated as follows:

$$\begin{aligned} E_z^{n+1}(2, n_y+1, k) = {} & [w_1 E_z^{n+1}(2+n_x-\left\lceil \tfrac{S_x}{\Delta x}\right\rceil, 1, k) \\ & + w_2 E_z^{n+1}(3+n_x-\left\lceil \tfrac{S_x}{\Delta x}\right\rceil, 1, k)]\times e^{-jk_x(S_x-P_x)}\times e^{-jk_y P_y}. \end{aligned} \tag{3.16}$$

If $n_x = 5$ in Equation (3.16), then for this specific case ceil $(S_x/\Delta x)$ will equal 2. The component $E_z^{n+1}(3+n_x-\left\lceil \tfrac{S_x}{\Delta x}\right\rceil, 1, k) = E_z^{n+1}(6,1,k)$ is a corner component so for the proper updating sequence, the corner E_z components ($y = 0$) should be updated before updating the E_z on the

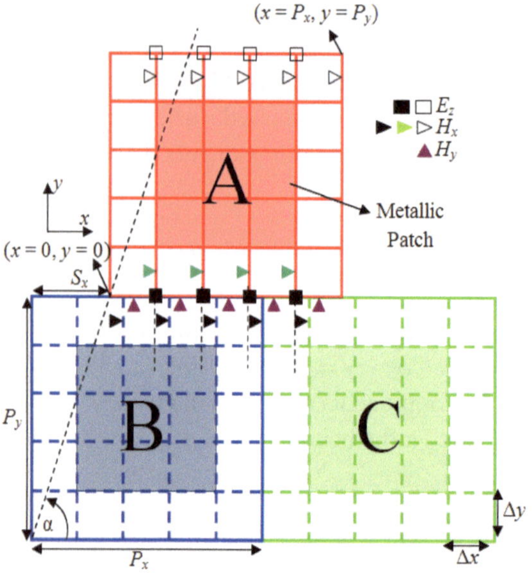

Figure 3.6: FDTD grid for non-coincident case, E_z components.

boundary $y = P_y$. Similar to the component $E_z^{n+1}(2, n_y + 1, k)$, all other E_z components on the boundary $y = P_y$ can be updated.

The E_z components at the corners are updated according to Fig. 3.7 as follows:

At $x = 0$ and $y = 0$

$$\begin{aligned} E_z^{n+1}(1,1,k) = C_{eze}(1,1,k) \times E_z^n(1,1,k) \\ + C_{ezhy}(1,1,k) \times [H_y^{n+1/2}(1,1,k) - H_y^{n+1/2}(n_x,1,k) \times e^{jk_x P_x}] \\ + C_{ezhx}(1,1,k) \times [H_x^{n+1/2}(1,1,k) - H_x^{n+1/2}(1,0,k)], \end{aligned} \tag{3.17}$$

where

$$H_x^{n+\frac{1}{2}}(1,0,k) = [w_1 H_x^{n+\frac{1}{2}}(1 + \left\lceil \frac{S_x}{\Delta x} \right\rceil, n_y, k) + w_2 H_x^{n+\frac{1}{2}}(\left\lceil \frac{S_x}{\Delta x} \right\rceil, n_y, k)] \times e^{jk_x S_x} \times e^{jk_y P_y}. \tag{3.18}$$

At $x = P_x$ and $y = 0$

$$E_z^{n+1}(n_x + 1, 1, k) = E_z^{n+1}(1,1,k) \times e^{-jk_x P_x}. \tag{3.19}$$

At $x = 0$ and $y = P_y$

$$\begin{aligned} E_z^{n+1}(1, n_y + 1, k) = \Bigg[w_1 E_z^{n+1}(1 + n_x - \left\lceil \frac{S_x}{dx} \right\rceil, 1, k) \\ + w_2 E_z^{n+1}(2 + n_x - \left\lceil \frac{S_x}{dx} \right\rceil, 1, k) \Bigg] \times e^{-jk_x(S_x - P_x)} \times e^{-jk_y P_y}. \end{aligned} \tag{3.20}$$

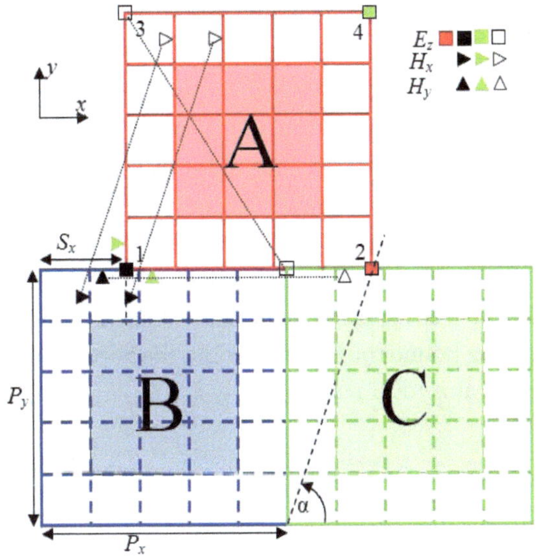

Figure 3.7: FDTD grid for non-coincident case, E_z corner components.

At $x = P_x$ and $y = P_y$

$$E_z^{n+1}\left(n_x+1, n_y+1, k\right) = E_z^{n+1}(1, n_y+1, k) \times e^{-jk_x P_x}. \tag{3.21}$$

The above procedure provides the six FDTD updating equations for the case of the non-coincident skewed shift. These updating equations can be used to update the electric and magnetic fields in any region in the computational domain (boundary and non-boundary).

3.3 NUMERICAL RESULTS

In this section, numerical results generated using the new algorithm are presented. The FDTD code was developed using MATLAB [19]. All the test cases were executed using the same computer (Intel Core 2 CPU 6700 2.66 GHz with 2 GB RAM). These results demonstrate the validity of the new algorithm for determining reflection and transmission properties of periodic structures with arbitrary skewed grids. The first example is an infinite dielectric slab excited by TM^z and TE^z plane waves. The second example is a dipole FSS, where the structure is analyzed with special skewed angles that can be simulated using the normal FDTD/PBC, and the third example is a JC FSS. The results obtained from the skewed FDTD code are compared with results obtained from an analytic solution, the axial FDTD method, and Ansoft Designer.

3.3.1 AN INFINITE DIELECTRIC SLAB

Due to its homogeneity, the infinite dielectric slab can be considered as a periodic structure with any skew angle. The algorithm is first used to analyze an infinite dielectric slab with thickness h = 9.375 mm and relative permittivity ε_r = 2.56. The slab is illuminated by TM^z and TE^z plane waves, respectively. The skew angle of the slab is set to 60°. The slab is excited using a cosine-modulated Gaussian pulse centered at 10 GHz with 20 GHz bandwidth. The plane wave is incident normally ($k_x = k_y = 0$ m^{-1}) and obliquely ($k_x = 104.8$ m^{-1}, $k_y = 0$ m^{-1} for minimum frequency of 5 GHz). The FDTD grid cell size is $\Delta x = \Delta y = \Delta z = 0.3125$ mm, and the slab is represented by 5x5 cells. In the FDTD code, 2,500 time steps and 0.9 reduction factor of the CFL time step are used. The CPML is used for the absorbing boundaries at the top and the bottom of the computational domain. The results are compared with analytical results in Fig. 3.8.

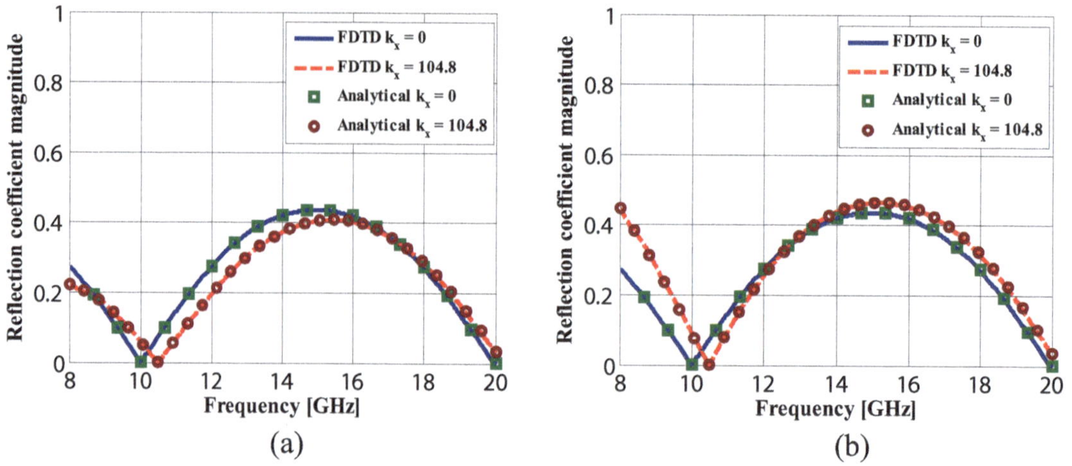

Figure 3.8: Reflection coefficient for infinite dielectric slab, (a) TM^z case, (b) TE^z case. (From [25] © IEEE).

From Fig. 3.8, good agreement between results based on the analytical solution and results generated by the new algorithm for both TM^z and TE^z cases (normal and oblique incidence) can be noticed. The stability of the algorithm is observed even at the angles of incidence near grazing.

3.3.2 A DIPOLE FSS

The algorithm is then used to analyze an FSS structure consisting of dipole elements. The dipole length is 12 mm and its width is 3 mm. The periodicity is 15 mm in both $x-$ and $y-$ directions. The substrate has a thickness of 6 mm and relative permittivity ε_r = 2.2, as shown in Fig. 3.9. The structure is first illuminated by a normally incident plane wave (with polarization along the y-axis), and the skew angle of the structure is set to 90° (axial case) and 63.43° (special case where the shift is a half unit cell in $x-$ direction). These two cases are special cases that can also be simulated using the axial periodic boundary conditions.

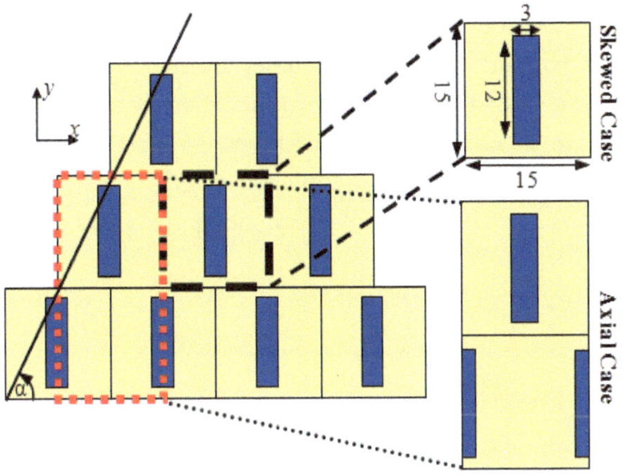

Figure 3.9: Dipole FSS geometry with skew angle $\alpha = 63.43°$ (all dimensions are in mm).

Figure 3.10 provides results for normal incidence. The structure is excited using a cosine-modulated Gaussian pulse centered at 8 GHz with 16 GHz bandwidth. In the FDTD code, 2,500 time steps and 0.9 reduction factor of the CFL time step are used. The CPML is used for the

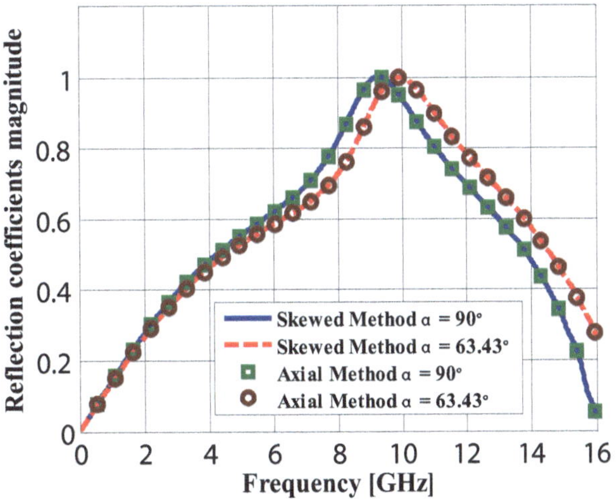

Figure 3.10: Reflection coefficient for a dipole FSS under normal incident TE^z plane wave with skew angle of 90° and 63.43°. (From [25] © IEEE).

absorbing boundaries at the top and the bottom of the computational domain. The FDTD grid cell size is $\Delta x = \Delta y = \Delta z = 0.5$ mm. The results are compared with results obtained from the axial FDTD code. The computational time per simulation for the skewed code is 4.28 minutes, and the memory usage is 0.2 MB. For the axial code with $\alpha = 63.43°$, the time is doubled due to the increase in the computational domain size (the unit cell size is doubled in the y-direction as shown in Fig. 3.9). A good agreement is observed between the results from the skewed method and from the axial method.

Figure 3.11 provides results for an oblique incidence ($\theta = 30°$ and $\phi = 60°$) exciting the dipole FSS structure with skew angle $\alpha = 50°$ (a general skewed grid which can't be implemented using the axial FDTD). To generate results for many frequency points and a specific angle of incidence,

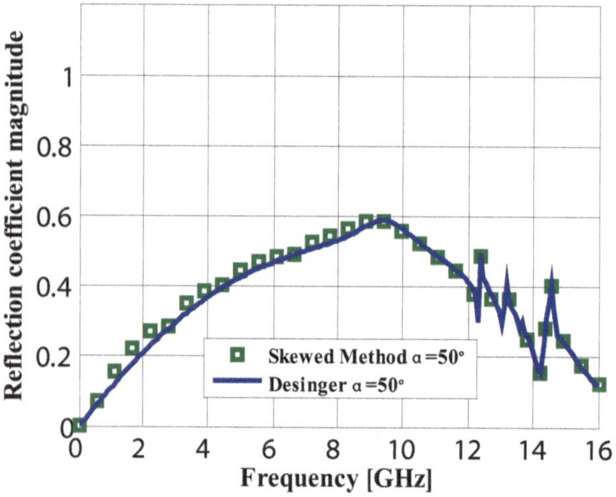

Figure 3.11: Reflection coefficient for a dipole FSS under oblique incident TE^z plane wave ($\theta = 30°$, $\phi = 60°$) with skew angle of 50°. (From [25] © IEEE).

multiple runs of the code are needed. The results in Fig. 3.11 are generated using 33 different k_x values (from 0.131 m^{-1} to 83.834 m^{-1}). The results are compared with results obtained from Ansoft Designer. From Fig. 3.11, a good agreement between the results generated using Ansoft Designer and those generated using the new algorithm can be noticed for this oblique incident case.

3.3.3 A JERUSALEM CROSS FSS

Next, the algorithm is used to analyze an FSS structure consisting of JC elements. The periodicity is 15.2 mm in both the $x-$ and y-directions. The dimensions of the elements are shown in Fig. 3.12. The structure is illuminated by a TE^z plane wave (polarization along y-axis). Figure 3.13 provides results for normal incidence. The structure is excited using a cosine-modulated Gaussian pulse centered at 7 GHz with 8 GHz bandwidth. The grid cell size is $\Delta x = \Delta y = 0.2285$ mm and $\Delta z =$ 0.457 mm.

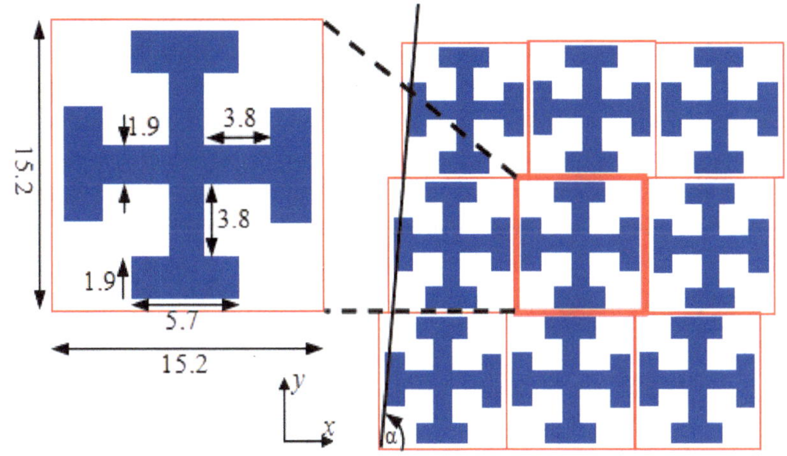

Figure 3.12: JC FSS geometry with skew angle $\alpha = 80°$ (all dimensions are in mm).

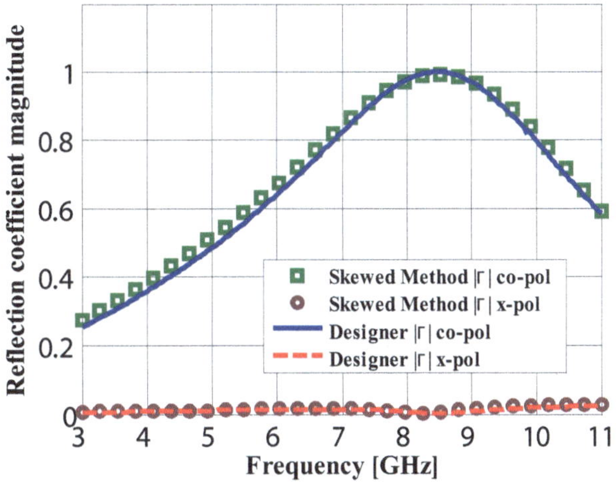

Figure 3.13: Co- and cross-polarization reflection coefficient for a JC FSS with skew angle of 80° under a normal incident TE^z plane wave. (From [26] © IEEE).

In the FDTD code, 3,000 time steps and a 0.9 reduction factor of the CFL time step are used. CPML is used for the absorbing boundaries at the top and the bottom. The structure has a skew angle α = 80° (general skewed grid). The results were compared with results obtained from Ansoft Designer. The computational time per simulation for the skewed code is 4.53 minutes, and the memory usage is 0.2 MB, while for Ansoft Designer computational time, per simulation is 45 minutes for 30 frequency points, and the memory usage is 21 MB using the same computer.

Figure 3.14 provides results for an oblique incidence plane wave ($\theta = 60°$ and $\phi = 45°$) exciting a JC FSS structure with skew angle $\alpha = 80°$. To generate results for many frequency points and a specific angle of incidence, multiple runs of the code are needed, which increases the computational time. Using 30 different k_x values (from 38.5031 m^{-1} to 141.1781 m^{-1}), both co- and cross-polarization reflection coefficients were generated. The results were compared with results obtained from Ansoft Designer. Good agreement between the results generated using Ansoft Designer and results generated using the new algorithm can be noticed in Figs. 3.13 and 3.14 for both normal and oblique incidences.

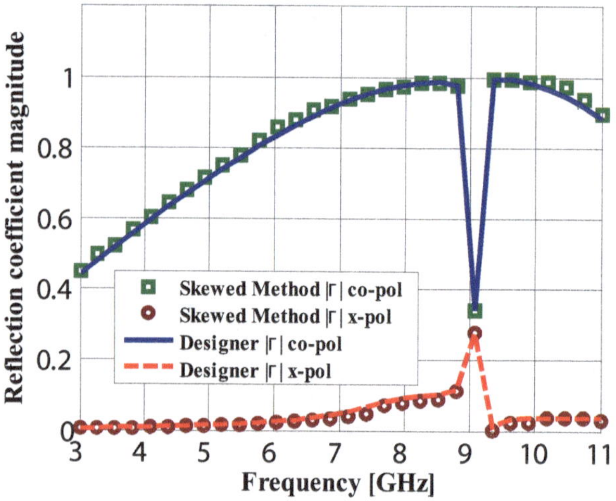

Figure 3.14: Co- and cross-polarization reflection coefficient for a JC FSS with skew angle $\alpha = 80°$ under an oblique incident TE^z plane wave ($\theta = 60°$, $\phi = 45°$). (From [26] © IEEE).

3.4 SUMMARY

This chapter introduces a new FDTD approach to analyze the scattering properties of general skewed grid periodic structures. The approach is developed based on the constant horizontal wavenumber technique. It is simple to implement and efficient in terms of both computational time and memory usage. In addition, the stability criterion is angle-independent. It is capable of calculating the co- and cross-polarized reflection and transmission coefficients of normal and oblique incidences, for both TE^z and TM^z cases, and for arbitrary skewed angles in both cases coincident and non-coincident shift. The numerical results show good agreement with results from the analytical solution for a dielectric slab, and with results based on the MoM solutions for dipole and JC FSS structures.

CHAPTER 4

Dispersive Periodic Structures

4.1 INTRODUCTION

Electromagnetic simulation of dispersive media is essential in many applications such as medical telemetries, metamaterials designs, nanoplasmonic solar cells, shielding materials electromagnetic compatibility, etc. Debye media, Lorentz media, and Drude media are three important classes of dispersive materials and reflect different frequency-dependent behaviors of the materials. Various FDTD formulations have been developed to simulate these frequency-dependent materials. The recursive convolution (RC) method [27, 28, 29, 30, 31, 32] and the auxiliary differential equation (ADE) method [33, 34] are the two very well known approaches. Piecewise linear recursive convolution [35] and the Z-transform [36, 37, 38] are also used to model dispersive media.

It's worthwhile to point out that most PBCs for the FDTD technique were developed to analyze periodic structures with dispersive media not extended to the boundary of the unit cells. However, there are numerous applications where periodic structures with dispersive media on the boundaries of the unit cell must be used. In this chapter, a new dispersive periodic boundary condition (DPBC) for the FDTD technique is developed to solve the above challenging problem. The algorithm utilizes the ADE technique with a two-term Debye relaxation equation to simulate the general dispersive property in the medium. In addition, the constant horizontal wavenumber approach is modified accordingly to implement the periodic boundary conditions. The new algorithm offers many advantages such as implementation simplicity, stability, and computational efficiency similar to the conventional FDTD characteristics.

The chapter is organized as follows: In Section 4.2, the description of ADE technique is provided. In Section 4.3, the FDTD updating equations are derived and the DPBC is described. In Section 4.4, several numerical examples validating the new approach are presented, including an infinite dispersive slab, nanoplasmonic solar cells, and a sandwiched composite frequency selective surface (FSS) structure. Various incident angles and polarizations have been tested in these examples, and the numerical results show good agreement with the analytical results or other numerical results obtained from frequency-domain methods. Section 4.5 provides a summary of this work.

4.2 AUXILIARY DIFFERENTIAL EQUATION METHOD

In the auxiliary differential equation (ADE) method, a differential equation relating the electric displacement vector $\boldsymbol{D}$ to the electric field vector $\boldsymbol{E}$ is added to the FDTD updating procedure. Solving this new equation simultaneously with the standard FDTD equations will lead to simulating the dispersive property of the medium [39, 40, 41]. The time-domain Maxwell's equations can be stated as in (2.1). For dispersive material the electric displacement vector and the magnetic flux density vector are described as:

$$\vec{D} = \varepsilon(\omega)\vec{E}, \tag{4.1a}$$
$$\vec{B} = \mu(\omega)\vec{H}, \tag{4.1b}$$

with frequency-dependent complex permeability and permittivity. Assuming in this book that only ε depends on the frequency, then (4.1b) can be written as $\vec{B} = \mu\vec{H}$. The dispersive characteristics of $\varepsilon(\omega)$ can be described by a two-term Debye relaxation equation as

$$\varepsilon(\omega) = \varepsilon_o\left[\varepsilon_\infty + \frac{\varepsilon_{s1} - \varepsilon_\infty}{1 + j\omega\tau_1} + \frac{\varepsilon_{s2} - \varepsilon_\infty}{1 + j\omega\tau_2}\right], \tag{4.2}$$

where ε_o is the free space permittivity, ε_s is the static or zero frequency relative permittivity, ε_∞ is the relative permittivity at infinite frequency and τ is the relaxation time. From (4.1a) and (4.2) $\vec{D}(\omega)$ can be written as follows:

$$\vec{D}(\omega) = \varepsilon_o\frac{\varepsilon_s + j\omega(\varepsilon_{s1}\tau_2 + \varepsilon_{s2}\tau_1) - \omega^2\tau_1\tau_2\varepsilon_\infty}{1 + j\omega(\tau_1 + \tau_2) - \omega^2\tau_1\tau_2}\vec{E}(\omega), \tag{4.3}$$

where the zero (static) frequency dielectric constant ε_s is given by

$$\varepsilon_s = \varepsilon_{s1} + \varepsilon_{s2} - \varepsilon_\infty. \tag{4.4}$$

From time harmonic expression in (4.3) to differential time-domain form using the relations $j\omega \rightarrow \frac{\partial}{\partial t}$, $-\omega^2 \rightarrow \frac{\partial^2}{\partial t^2}$ one can re-write (4.3) as:

$$\vec{D}(t)+(\tau_1+\tau_2)\frac{\partial\vec{D}(t)}{\partial t}+\tau_1\tau_2\frac{\partial^2\vec{D}(t)}{\partial t^2}=\varepsilon_o\varepsilon_s\vec{E}(t) + \varepsilon_o(\varepsilon_{s1}\tau_2+\varepsilon_{s2}\tau_1)\frac{\partial\vec{E}(t)}{\partial t} + \tau_1\tau_2\varepsilon_o\varepsilon_\infty\frac{\partial^2\vec{E}(t)}{\partial t^2}. \tag{4.5}$$

Using Equation (2.1a), (2.1b), and (4.5), each vector equation can be decomposed to three scalar equations in a three-dimensional Cartesian space. Therefore, Maxwell's curl equations can be represented with nine scalar equations in the Cartesian coordinate system (x, y, z) relating $\boldsymbol{H}$ to $\boldsymbol{E}$ and $\boldsymbol{E}$ to $\boldsymbol{D}$ as follows with the conduction and displacement currents combined in the definition of the complex permittivity $\varepsilon(\omega)$:

$$\frac{\partial H_x}{\partial t} = \frac{1}{\mu_x}\left[\frac{\partial E_y}{\partial z} - \frac{\partial E_z}{\partial y} - \sigma_x^m H_x - M_{ix}\right], \tag{4.6a}$$

$$\frac{\partial H_y}{\partial t} = \frac{1}{\mu_y}\left[\frac{\partial E_z}{\partial x} - \frac{\partial E_x}{\partial z} - \sigma_y^m H_y - M_{iy}\right], \tag{4.6b}$$

$$\frac{\partial H_z}{\partial t} = \frac{1}{\mu_z}\left[\frac{\partial E_x}{\partial y} - \frac{\partial E_y}{\partial x} - \sigma_z^m H_z - M_{iz}\right] \tag{4.6c}$$

$$\frac{\partial D_x}{\partial t} = \left[\frac{\partial H_z}{\partial y} - \frac{\partial H_y}{\partial z} - J_{ix}\right], \tag{4.6d}$$

$$\frac{\partial D_y}{\partial t} = \left[\frac{\partial H_x}{\partial z} - \frac{\partial H_z}{\partial x} - J_{iy}\right], \tag{4.6e}$$

$$\frac{\partial D_z}{\partial t} = \left[\frac{\partial H_y}{\partial x} - \frac{\partial H_x}{\partial y} - J_{iz}\right], \tag{4.6f}$$

$$D_x + (\tau_1+\tau_2)\frac{\partial D_x}{\partial t} + \tau_1\tau_2\frac{\partial^2 D_x}{\partial t^2} = \varepsilon_o\varepsilon_s E_x + \varepsilon_o(\varepsilon_{s1}\tau_2 + \varepsilon_{s2}\tau_1)\frac{\partial E_x}{\partial t} + \tau_1\tau_2\varepsilon_o\varepsilon_\infty\frac{\partial^2 E_x}{\partial t^2}, \tag{4.6g}$$

$$D_y + (\tau_1+\tau_2)\frac{\partial D_y}{\partial t} + \tau_1\tau_2\frac{\partial^2 D_y}{\partial t^2} = \varepsilon_o\varepsilon_s E_y + \varepsilon_o(\varepsilon_{s1}\tau_2 + \varepsilon_{s2}\tau_1)\frac{\partial E_y}{\partial t} + \tau_1\tau_2\varepsilon_o\varepsilon_\infty\frac{\partial^2 E_y}{\partial t^2}, \tag{4.6h}$$

$$D_z + (\tau_1+\tau_2)\frac{\partial D_z}{\partial t} + \tau_1\tau_2\frac{\partial^2 D_z}{\partial t^2} = \varepsilon_o\varepsilon_s E_z + \varepsilon_o(\varepsilon_{s1}\tau_2 + \varepsilon_{s2}\tau_1)\frac{\partial E_z}{\partial t} + \tau_1\tau_2\varepsilon_o\varepsilon_\infty\frac{\partial^2 E_z}{\partial t^2}. \tag{4.6i}$$

Re-arranging the above nine equations, the iterative FDTD simulation can be easily constructed. As long as the μ (permeability) of the material is independent of frequency, the last updating equations for the magnetic field will be similar to those in the conventional FDTD formulation.

1) First, to obtain the updating equations of the electric displacement vector $\boldsymbol{D}$, we start by updating D_x as follows:

$$\frac{D_x^{n+1}(i,j,k) - D_x^n(i,j,k)}{\Delta t} = [\frac{H_z^{n+\frac{1}{2}}(i,j,k) - H_z^{n+\frac{1}{2}}(i,j-1,k)}{\Delta y} - \frac{H_y^{n+\frac{1}{2}}(i,j,k) - H_y^{n+\frac{1}{2}}(i,j,k-1)}{\Delta z} - J_{ix}^{n+\frac{1}{2}}(i,j,k)],$$

$$\begin{aligned} D_x^{n+1}(i,j,k) = {} & C_{dxd}(i,j,k) \times D_x^n(i,j,k) + C_{dxhz}(i,j,k) \times [H_z^{n+\frac{1}{2}}(i,j,k) \\ & - H_z^{n+\frac{1}{2}}(i,j-1,k)] + C_{dxhy}(i,j,k) \times [H_y^{n+\frac{1}{2}}(i,j,k) - H_y^{n+\frac{1}{2}}(i,j,k-1)] \\ & + C_{dxj}(i,j,k) \times J_{ix}^{n+\frac{1}{2}}(i,j,k), \end{aligned} \tag{4.7}$$

where

$$C_{dxd}(i,j,k) = 1,\; C_{dxhz}(i,j,k) = \frac{\Delta t}{\Delta y},\; C_{dxhy}(i,j,k) = \frac{-\Delta t}{\Delta z},\; C_{dxj}(i,j,k) = -\Delta t,$$

similarly for D_y:

$$D_y^{n+1}(i,j,k) = C_{dyd}(i,j,k) \times D_y^n(i,j,k) + C_{dyhx}(i,j,k) \times [H_x^{n+\frac{1}{2}}(i,j,k) - H_x^{n+\frac{1}{2}}(i,j,k-1)] + C_{dyhz}(i,j,k) \times [H_z^{n+\frac{1}{2}}(i,j,k) - H_z^{n+\frac{1}{2}}(i-1,j,k)] + C_{dyj}(i,j,k) \times J_{iy}^{n+\frac{1}{2}}(i,j,k), \tag{4.8}$$

where

$$C_{dyd}(i,j,k) = 1,\ C_{dyhx}(i,j,k) = \frac{\Delta t}{\Delta z},\ C_{dyhz}(i,j,k) = \frac{-\Delta t}{\Delta x},\ C_{dyj}(i,j,k) = -\Delta t,$$

and for D_z:

$$D_z^{n+1}(i,j,k) = C_{dzd}(i,j,k) \times D_z^n(i,j,k) + C_{dzhy}(i,j,k) \times [H_y^{n+\frac{1}{2}}(i,j,k) - H_y^{n+\frac{1}{2}}(i-1,j,k)] + C_{dzhx}(i,j,k) \times [H_x^{n+\frac{1}{2}}(i,j,k) - H_x^{n+\frac{1}{2}}(i,j-1,k)] + C_{dzj}(i,j,k) \times J_{iz}^{n+\frac{1}{2}}(i,j,k), \tag{4.9}$$

where

$$C_{dzd}(i,j,k) = 1,\ C_{dzhy}(i,j,k) = \frac{\Delta t}{\Delta x},\ C_{dzhx}(i,j,k) = \frac{-\Delta t}{\Delta y},\ C_{dzj}(i,j,k) = -\Delta t.$$

2) To obtain the updating equations for the electric field vector $\boldsymbol{E}$, one can start by updating E_x as follows:

$$D_x + (\tau_1^x + \tau_2^x)\frac{\partial D_x}{\partial t} + \tau_1^x \tau_2^x \frac{\partial^2 D_x}{\partial t^2} = \varepsilon_o \varepsilon_s^x E_x + \varepsilon_o(\varepsilon_{s1}^x \tau_2^x + \varepsilon_{s2}^x \tau_1^x)\frac{\partial E_x}{\partial t} + \tau_1^x \tau_2^x \varepsilon_o \varepsilon_\infty^x \frac{\partial^2 E_x}{\partial t^2}.$$

Using central differences centered at time step $(n + {}^1\!/\!_2)$ the above equation can be written as

$$\frac{D_x^{n+1} + D_x^n}{2} + (\tau_1^x + \tau_2^x)\frac{D_x^{n+1} - D_x^n}{\Delta t} + \tau_1^x \tau_2^x \frac{D_x^{n+1} - 2D_x^n + D_x^{n-1}}{(\Delta t)^2} = \varepsilon_o \varepsilon_s^x \frac{E_x^{n+1} + E_x^n}{2} + \varepsilon_o(\varepsilon_{s1}^x \tau_2^x + \varepsilon_{s2}^x \tau_1^x)\frac{E_x^{n+1} - E_x^n}{\Delta t} + \tau_1^x \tau_2^x \varepsilon_o \varepsilon_\infty^x \frac{E_x^{n+1} - 2E_x^n + E_x^{n-1}}{(\Delta t)^2}.$$

Taking $\beta_0^x = \frac{1}{2}$, $\beta_1^x = \frac{(\tau_1^x + \tau_2^x)}{\Delta t}$, $\beta_2^x = \frac{\tau_1^x \tau_2^x}{(\Delta t)^2}$, $\alpha_0^x = \frac{\varepsilon_o \varepsilon_s^x}{2}$, $\alpha_1^x = \frac{\varepsilon_o(\varepsilon_{s1}^x \tau_2^x + \varepsilon_{s2}^x \tau_1^x)}{\Delta t}$, and $\alpha_2^x = \frac{\tau_1^x \tau_2^x \varepsilon_o \varepsilon_\infty^x}{(\Delta t)^2}$ we have

$$\begin{aligned}
&\beta_0^x(D_x^{n+1} + D_x^n) + \beta_1^x(D_x^{n+1} - D_x^n) + \beta_2^x(D_x^{n+1} - 2D_x^n + D_x^{n-1}) = \\
&\alpha_0^x(E_x^{n+1} + E_x^n) + \alpha_1^x(E_x^{n+1} - E_x^n) + \alpha_2^x(E_x^{n+1} - 2E_x^n + E_x^{n-1}) \\
&[\alpha_0^x + \alpha_1^x + \alpha_2^x]E_x^{n+1} = [-\alpha_0^x + \alpha_1^x + 2\alpha_2^x]E_x^n + [-\alpha_2^x]E_x^{n-1} \\
&\quad + [\beta_0^x + \beta_1^x + \beta_2^x]D_x^{n+1} + [\beta_0^x - \beta_1^x - 2\beta_2^x]D_x^n + [\beta_2^x]D_x^{n-1},
\end{aligned}$$

which yields

$$E_x^{n+1}(i,j,k) = C_{exe1} \times E_x^n(i,j,k) + C_{exe2} \times E_x^{n-1}(i,j,k) + C_{exd1} \times D_x^{n+1}(i,j,k) + C_{exd2} \times D_x^n(i,j,k) + C_{exd3} \times D_x^{n-1}(i,j,k) \tag{4.10}$$

where

$$C_{exe1} = \frac{[-\alpha_0^x + \alpha_1^x + 2\alpha_2^x]}{[\alpha_0^x + \alpha_1^x + \alpha_2^x]}, \quad C_{exe2} = \frac{[-\alpha_2^x]}{[\alpha_0^x + \alpha_1^x + \alpha_2^x]}, \quad C_{exd1} = \frac{[\beta_0^x + \beta_1^x + \beta_2^x]}{[\alpha_0^x + \alpha_1^x + \alpha_2^x]},$$
$$C_{exd2} = \frac{[\beta_0^x - \beta_1^x - 2\beta_2^x]}{[\alpha_0^x + \alpha_1^x + \alpha_2^x]}, \quad C_{exd3} = \frac{[\beta_2^x]}{[\alpha_0^x + \alpha_1^x + \alpha_2^x]}.$$

Similarly, for E_y:

$$E_y^{n+1}(i,j,k) = C_{eye1} \times E_y^n(i,j,k) + C_{eye2} \times E_y^{n-1}(i,j,k) + C_{eyd1} \times D_y^{n+1}(i,j,k) + C_{eyd2} \times D_y^n(i,j,k) + C_{eyd3} \times D_y^{n-1}(i,j,k) \tag{4.11}$$

where

$$C_{eye1} = \frac{[-\alpha_0^y + \alpha_1^y + 2\alpha_2^y]}{[\alpha_0^y + \alpha_1^y + \alpha_2^y]}, \quad C_{eye2} = \frac{[-\alpha_2^y]}{[\alpha_0^y + \alpha_1^y + \alpha_2^y]}, \quad C_{eyd1} = \frac{[\beta_0^y + \beta_1^y + \beta_2^y]}{[\alpha_0^y + \alpha_1^y + \alpha_2^y]},$$
$$C_{eyd2} = \frac{[\beta_0^y - \beta_1^y - 2\beta_2^y]}{[\alpha_0^y + \alpha_1^y + \alpha_2^y]}, \quad C_{eyd3} = \frac{[\beta_2^y]}{[\alpha_0^y + \alpha_1^y + \alpha_2^y]},$$

and for E_z:

$$E_z^{n+1}(i,j,k) = C_{eze1} \times E_z^n(i,j,k) + C_{eze2} \times E_z^{n-1}(i,j,k) + C_{ezd1} \times D_z^{n+1}(i,j,k) + C_{ezd2} \times D_z^n(i,j,k) + C_{ezd3} \times D_z^{n-1}(i,j,k) \tag{4.12}$$

where

$$C_{eze1} = \frac{[-\alpha_0^z + \alpha_1^z + 2\alpha_2^z]}{[\alpha_0^z + \alpha_1^z + \alpha_2^z]}, \quad C_{eze2} = \frac{[-\alpha_2^z]}{[\alpha_0^z + \alpha_1^z + \alpha_2^z]}, \quad C_{ezd1} = \frac{[\beta_0^z + \beta_1^z + \beta_2^z]}{[\alpha_0^z + \alpha_1^z + \alpha_2^z]},$$
$$C_{ezd2} = \frac{[\beta_0^z - \beta_1^z - 2\beta_2^z]}{[\alpha_0^z + \alpha_1^z + \alpha_2^z]}, \quad C_{ezd3} = \frac{[\beta_2^y]}{[\alpha_0^z + \alpha_1^z + \alpha_2^z]}.$$

3) For the magnetic field components the traditional updating equations for H_x, H_y, and H_z can be used as follows:

For the H_x component:

$$\begin{aligned} H_x^{n+\frac{1}{2}}(i,j,k) = {} & C_{hxh}(i,j,k) \times H_x^{n-\frac{1}{2}}(i,j,k) \\ & + C_{hxey}(i,j,k) \times \left[E_y^n(i,j,k+1) - E_y^n(i,j,k)\right] \\ & + C_{hxez}(i,j,k) \times \left[E_z^n(i,j+1,k) - E_z^n(i,j,k)\right] \\ & + C_{hxm}(i,j,k) \times M_{ix}^n(i,j,k), \end{aligned} \tag{4.13}$$

where

$$C_{hxh}(i,j,k) = \frac{2\mu_x(i,j,k) - \Delta t \sigma_x^m(i,j,k)}{2\mu_x(i,j,k) + \Delta t \sigma_x^m(i,j,k)}, \quad C_{hxey}(i,j,k) = \frac{2\Delta t}{(2\mu_x(i,j,k) + \Delta t \sigma_x^m(i,j,k))\Delta z},$$
$$C_{hxez}(i,j,k) = \frac{-2\Delta t}{(2\mu_x(i,j,k) + \Delta t \sigma_x^m(i,j,k))\Delta y}, \quad C_{hxm}(i,j,k) = \frac{-2\Delta t}{2\mu_x(i,j,k) + \Delta t \sigma_x^m(i,j,k)}.$$

For the H_y component:

$$\begin{aligned} H_y^{n+\frac{1}{2}}(i,j,k) = {} & C_{hyh}(i,j,k) \times H_y^{n-\frac{1}{2}}(i,j,k) \\ & + C_{hyez}(i,j,k) \times \left[E_z^n(i+1,j,k) - E_z^n(i,j,k)\right] \\ & + C_{hyex}(i,j,k) \times \left[E_x^n(i,j,k+1) - E_x^n(i,j,k)\right] \\ & \ddot{y} + C_{hym}(i,j,k) \times M_{iy}^n(i,j,k), \end{aligned} \tag{4.14}$$

where

$$C_{hyh}(i,j,k) = \frac{2\mu_y(i,j,k) - \Delta t \sigma_y^m(i,j,k)}{2\mu_y(i,j,k) + \Delta t \sigma_y^m(i,j,k)}, \quad C_{hyez}(i,j,k) = \frac{2\Delta t}{(2\mu_y(i,j,k) + \Delta t \sigma_y^m(i,j,k))\Delta x},$$
$$C_{hyex}(i,j,k) = \frac{-2\Delta t}{(2\mu_y(i,j,k) + \Delta t \sigma_y^m(i,j,k))\Delta y}, \quad C_{hym}(i,j,k) = \frac{-2\Delta t}{2\mu_y(i,j,k) + \Delta t \sigma_y^m(i,j,k)},$$

and for the H_z component:

$$\begin{aligned} H_z^{n+\frac{1}{2}}(i,j,k) = {} & C_{hzh}(i,j,k) \times H_z^{n-\frac{1}{2}}(i,j,k) \\ & + C_{hzex}(i,j,k) \times \left[E_x^n(i,j+1,k) - E_x^n(i,j,k)\right] \\ & + C_{hxey}(i,j,k) \times \left[E_y^n(i+1,j,k) - E_y^n(i,j,k)\right] \\ & + C_{hzm}(i,j,k) \times M_{iz}^n(i,j,k), \end{aligned} \tag{4.15}$$

where

$$C_{hzh}(i,j,k) = \frac{2\mu_z(i,j,k) - \Delta t \sigma_z^m(i,j,k)}{2\mu_z(i,j,k) + \Delta t \sigma_z^m(i,j,k)}, \quad C_{hzex}(i,j,k) = \frac{2\Delta t}{(2\mu_z(i,j,k) + \Delta t \sigma_z^m(i,j,k))\Delta y},$$
$$C_{hzey}(i,j,k) = \frac{-2\Delta t}{(2\mu_z(i,j,k) + \Delta t \sigma_z^m(i,j,k))\Delta x}, \quad C_{hzm}(i,j,k) = \frac{-2\Delta t}{2\mu_z(i,j,k) + \Delta t \sigma_z^m(i,j,k)}.$$

Using Equations (4.7)–(4.15), a complete FDTD algorithm for dispersive materials with a frequency-dependent permittivity is constructed. As in the conventional (frequency-independent) FDTD method, the fields $\boldsymbol{E}$ and $\boldsymbol{H}$ are calculated in a time-stepping manner for a lattice of Yee cells. In this formulation the values of $\boldsymbol{E}$ are used to calculate $\boldsymbol{H}$ from (4.13), (4.14), and (4.15); the values of $\boldsymbol{H}$ are used to calculate $\boldsymbol{D}$ from (4.7), (4.8), and (4.9); and the values of $\boldsymbol{D}$ are used to calculate $\boldsymbol{E}$ from (4.10), (4.11), and (4.12), after which the process is repeated iteratively.

The dispersive material equation and the developed code should also be capable of implementing normal dielectric material. This can easily be done by substituting zeros for both τ_1 and

τ_2 in Equation (4.2), hence the equation will be reduced to $\varepsilon(\omega) = \varepsilon_o\varepsilon_s$, where ε_s is given by (4.4). By substituting zeros for τ_1 and τ_2 in Equation (4.10) the parameters $\alpha_1 = \alpha_2 = \beta_1 = \beta_2 = 0$ and hence, the equation will be reduced to

$$E_x^{n+1}(i, j, k) = \frac{1}{\varepsilon_x(i, j, k)} D_x^{n+1}(i, j, k). \tag{4.16}$$

Equation (4.16) verifies that the material is a non-dispersive dielectric media. The y- and z-components can be also treated similarly. For an FDTD scattered field formulation for dispersive media the updating equations are listed in Appendix A.1.

4.3 DISPERSIVE PERIODIC BOUNDARY CONDITIONS

In this section a new DPBC is developed to analyze periodic structures with dispersive media extended to the boundaries of the unit cell. The new algorithm utilizes the conventional ADE technique to update the magnetic field components and the non-boundary electric field components. In addition, a modified version of the constant horizontal wavenumber approach is derived to update electric field components on the boundaries. The updated version of the constant horizontal wavenumber approach is based on Floquet analysis for the electrical field vector as follows:

$$E(x = 0, y, z, \omega) = E(x = P_x, y, z, \omega) \times e^{jk_x P_x}. \tag{4.17}$$

Multiplying both sides of Equation (4.17) by the complex permittivity will result in the following equation:

$$\varepsilon(\omega)E(x = 0, y, z, \omega) = \varepsilon(\omega)E(x = P_x, y, z, \omega) \times e^{jk_x P_x}. \tag{4.18}$$

This can be represented as follows:

$$D(x = 0, y, z, \omega) = D(x = P_x, y, z, \omega) \times e^{jk_x P_x}. \tag{4.19}$$

Equation (4.19) represents the Floquet theory for the displacement electric field vector $\boldsymbol{D}$. Using the constant horizontal wavenumber approach, Equation (4.19) can be directly transformed to the time-domain as follows:

$$D(x = 0, y, z, t) = D(x = P_x, y, z, t) \times e^{jk_x P_x}. \tag{4.20}$$

Using Equation (4.20) and the ADE technique, the updating equations for a periodic structure can be easily derived. The magnetic field components are updated using the FDTD updating Equations (4.13)–(4.15). The electric field components are updated using ADE FDTD updating Equations (4.10)–(4.12). While the non-boundary components of the electric displacement field vectors are updated using the ADE FDTD updating Equations (4.7)–(4.9). The components on the boundaries will be updated using DPBC equations based on the constant horizontal wavenumber approach. The updating equations for the boundary electric displacement field components are organized as follows:

1) Updating D_x at $y = 0$ and $y = P_y$.

2) Updating D_y at $x = 0$ and $x = P_x$.

3) Updating D_z at $y = 0$, $y = P_y$, $x = 0$, and $x = P_x$, without the corners.

4) Updating D_z at the corners.

1) To update the D_x on the boundary $y = 0$, the magnetic field components H_z outside the computational domain are needed as shown in Fig. 2.6. However, due to periodicity in the y-direction, one can use magnetic field components H_z of interest inside the computational domains; to update these electric displacement field vectors a procedure similar to the procedure in Section 2.3 is used. Starting from

$$\begin{aligned} D_x^{n+1}(i,1,k) = &C_{dxd}(i,1,k) D_x^n(i,1,k) + C_{dxhz}(i,1,k)[H_z^{n+1/2}(i,1,k) \\ &- H_z^{n+1/2}(i,0,k)] \\ &+ C_{dxhy}(i,1,k)[H_y^{n+1/2}(i,1,k) - H_y^{n+1/2}(i,1,k-1)], \end{aligned} \tag{4.21}$$

where the coefficients are stated as in (4.7), and $H_z^{n+1/2}(i,0,k) = H_z^{n+1/2}(i,n_y,k) \times e^{jk_y P_y}$ due to the periodicity in the y-direction as shown in Fig. 2.6. The term $e^{jk_y P_y}$ is used to compensate for the phase shift due to general oblique incidence. Then the updating equation for D_x on the boundary $y = 0$ can be written as

$$\begin{aligned} D_x^{n+1}(i,1,k) = &C_{dxd}(i,1,k) \times D_x^n(i,1,k) \\ &+ C_{dxhz}(i,1,k) \times [H_z^{n+1/2}(i,1,k) - H_z^{n+1/2}(i,n_y,k) \times e^{jk_y P_y}] \\ &+ C_{dxhy}(i,1,k) \times [H_y^{n+1/2}(i,1,k) - H_y^{n+1/2}(i,1,k-1)]. \end{aligned} \tag{4.22}$$

For D_x on the boundary $y = P_y$ the updating equation can be written as

$$D_x^{n+1}(i,n_y+1,k) = D_x^{n+1}(i,1,k) \times e^{-jk_y P_y}. \tag{4.23}$$

2) Due to periodicity in x-direction as shown in Fig. 2.7, the updating equation for the D_y component on the boundary $x = 0$ can be written as

$$\begin{aligned} D_y^{n+1}(1,j,k) = &C_{dyd}(1,j,k) \times D_y^n(1,j,k) \\ &+ C_{dyhx}(1,j,k) \times [H_x^{n+1/2}(1,j,k) - H_x^{n+1/2}(1,j,k-1)] \\ &+ C_{dyhz}(1,j,k) \times [H_z^{n+1/2}(1,j,k) - H_z^{n+1/2}(n_x,j,k) \times e^{jk_x P_x}]. \end{aligned} \tag{4.24}$$

For D_y on the boundary $x = P_x$ the updating equation can be written as

$$D_y^{n+1}(n_x+1,j,k) = D_y^{n+1}(1,j,k) \times e^{-jk_x P_x}. \tag{4.25}$$

3) A similar procedure is used for updating the D_z components, but corner components are updated separately due to the presence of periodicity in both the x- and y-directions.

For D_z on the boundaries $x = 0$ and $x = P_x$, the updating equation can be written for $j \neq 1$ and $j \neq n_y + 1$ (avoiding the corners) as

$$\begin{aligned} D_z^{n+1}(1, j, k) = & C_{dzd}(1, j, k) \times D_z^n(1, j, k) \\ & + C_{dzhy}(1, j, k) \times [H_y^{n+\frac{1}{2}}(1, j, k) - H_y^{n+\frac{1}{2}}(n_x, j, k) \times e^{jk_x P_x}] \\ & + C_{dzhx}(1, j, k) \times [H_x^{n+\frac{1}{2}}(1, j, k) - H_x^{n+\frac{1}{2}}(1, j-1, k)]. \end{aligned} \quad (4.26)$$

$$D_z^{n+1}(n_x + 1, j, k) = D_z^{n+1}(1, j, k) \times e^{-jk_x P_x}. \quad (4.27)$$

The updating equation for the D_z components on the boundaries $y = 0$, and $y = P_y$ can be written for $i \neq 1$ and $i \neq n_x + 1$ (avoiding the corners) as

$$\begin{aligned} D_z^{n+1}(i, 1, k) = & C_{dzd}(i, 1, k) \times D_z^n(i, 1, k) \\ & + C_{dzhy}(i, 1, k) \times [H_y^{n+\frac{1}{2}}(i, 1, k) - H_y^{n+\frac{1}{2}}(i-1, 1, k)] \\ & + C_{dzhx}(i, 1, k) \times [H_x^{n+\frac{1}{2}}(i, 1, k) - H_x^{n+\frac{1}{2}}(i, n_y, k) \times e^{jk_y P_y}]. \end{aligned} \quad (4.28)$$

$$D_z^{n+1}(i, n_y + 1, k) = D_z^{n+1}(i, 1, k) \times e^{-jk_y P_y}. \quad (4.29)$$

4) The D_z components at the corners are updated as follows:
At $x = 0$ and $y = 0$

$$\begin{aligned} D_z^{n+1}(1, 1, k) = & C_{dzd}(1, 1, k) \times D_z^n(1, 1, k) \\ & + C_{dzhy}(1, 1, k) \times [H_y^{n+\frac{1}{2}}(1, 1, k) - H_y^{n+\frac{1}{2}}(n_x, 1, k) \times e^{jk_x P_x}] \\ & + C_{dzhx}(1, 1, k) \times [H_x^{n+\frac{1}{2}}(1, 1, k) - H_x^{n+\frac{1}{2}}(1, n_y, k) \times e^{jk_y P_y}]. \end{aligned} \quad (4.30)$$

At $x = P_x$ and $y = 0$:

$$D_z^{n+1}(n_x + 1, 1, k) = D_z^{n+1}(1, 1, k) \times e^{-jk_x P_x}. \quad (4.31)$$

At $x = 0$ and $y = P_y$:

$$D_z^{n+1}(1, n_y + 1, k) = D_z^{n+1}(1, 1, k) \times e^{-jk_y P_y}. \quad (4.32)$$

At $x = P_x$ and $y = P_y$:

$$D_z^{n+1}(n_x + 1, n_y + 1, k) = D_z^{n+1}(1, 1, k) \times e^{-jk_y P_y} \times e^{-jk_x P_x}. \quad (4.33)$$

Equations (4.21)–(4.33) together with Equations (4.13)–(4.15) and (4.10)–(4.12) describe the new FDTD/DPBC. After deriving the updating equations, a time marching algorithm can be constructed, as shown in Fig. 4.1. The main difference between this algorithm and the conventional FDTD algorithm is that the computational domain is divided into four main regions, as shown in

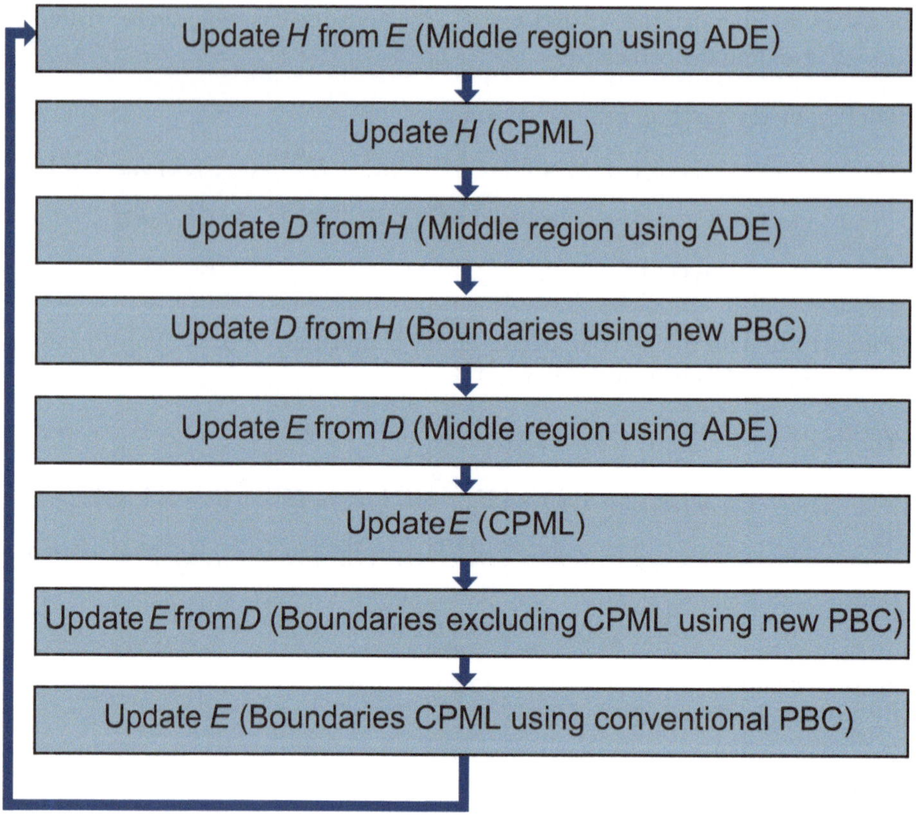

Figure 4.1: The flowchart of the new FDTD/DPBC code.

Fig. 4.2. The first region is the middle region where all components of $\boldsymbol{E}$, $\boldsymbol{H}$, and $\boldsymbol{D}$ are updated. The second region consists of the two CPML regions where only $\boldsymbol{E}$ and $\boldsymbol{H}$ are updated using the CPML (the CPML is not modified to handle dispersive media). The third region is the middle region of the boundaries where $\boldsymbol{D}$ is updated using the new DPBC. The fourth region consists of the boundaries of the CPML regions where only $\boldsymbol{E}$ is updated using the conventional PBC.

4.4 NUMERICAL RESULTS

In this section, numerical results generated using the new developed algorithm are presented. The FDTD code was developed in MATLAB and run on a computer with an Intel Core 2 CPU 6700, 2.66 GHz with 2 GB RAM. These results demonstrate the validity of the new algorithm for determining reflection and transmission properties of periodic structures with general dispersive media. The results generated by the new formulation are compared with results obtained from analytical so-

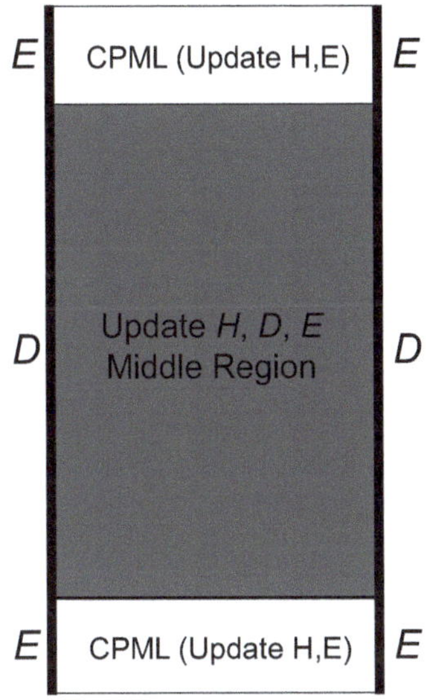

Figure 4.2: The four different regions of the new FDTD/DPBC computational domain.

lutions, the FDTD method with conventional PBC, and Ansoft high frequency structural simulator (HFSS), which is based on the finite element method (FEM) [43].

4.4.1 AN INFINITE WATER SLAB

The algorithm is first used to analyze an infinite water slab with thickness h = 6 mm. The slab is illuminated by TM^z and TE^z plane waves in two different simulations. The geometry of the slab is shown in Fig. 4.3. The parameters of water permittivity are obtained from [40] as $\varepsilon_{s1} = 81$, $\varepsilon_{s2} = 1.8$, $\varepsilon_\infty = 1.8$, $\tau_1 = 9.4 \times 10^{-12}$ and $\tau_2 = 0$. The permittivity of water versus frequency is shown in Fig. 4.4. The FDTD grid cell size is $\Delta x = \Delta y = \Delta z = 0.125$ mm, and the slab is represented by 2 $\times$ 2 cells (due to the homogeneity of the infinite slab it could be considered as a periodic structure with any periodicity). In the FDTD code 10,000 time steps and a 0.9 reduction factor of CFL time step are used. The CPML was used for the absorbing boundaries at the top and the bottom of the computational domain. The slab is excited using a cosine-modulated Gaussian pulse centered at 10 GHz with 20 GHz bandwidth for the normal incidence case (k_x = 0 m^{-1}), and it is excited using a cosine-modulated Gaussian pulse centered at 12.75 GHz with 14.5 GHz bandwidth for the oblique incidence case (k_x =104.8 m^{-1} for minimum frequency of 5 GHz) [44]. The results are compared with those obtained from the analytical formulations.

From Figs. 4.5 and 4.6 good agreement between analytical solutions and results generated by the new FDTD/DPBC algorithm can be noticed for both TM^z and TE^z cases (normal and oblique incidence). The computational time is equal to 1.17 minutes for each FDTD simulation.

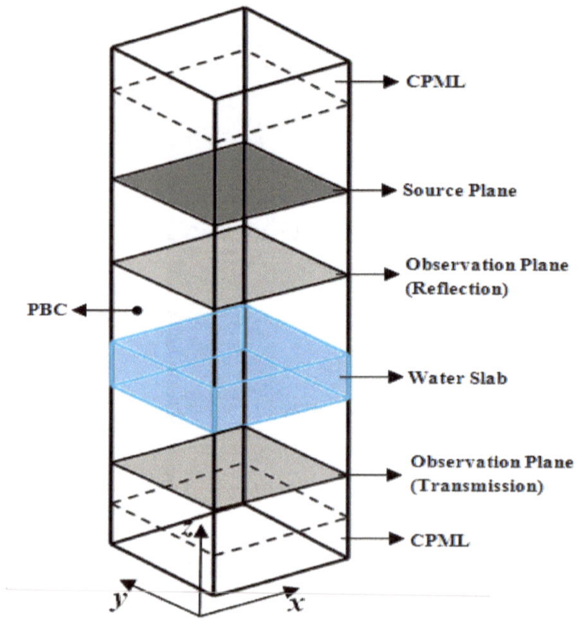

Figure 4.3: Geometry of the simulated infinite water slab (from [44] © IEEE).

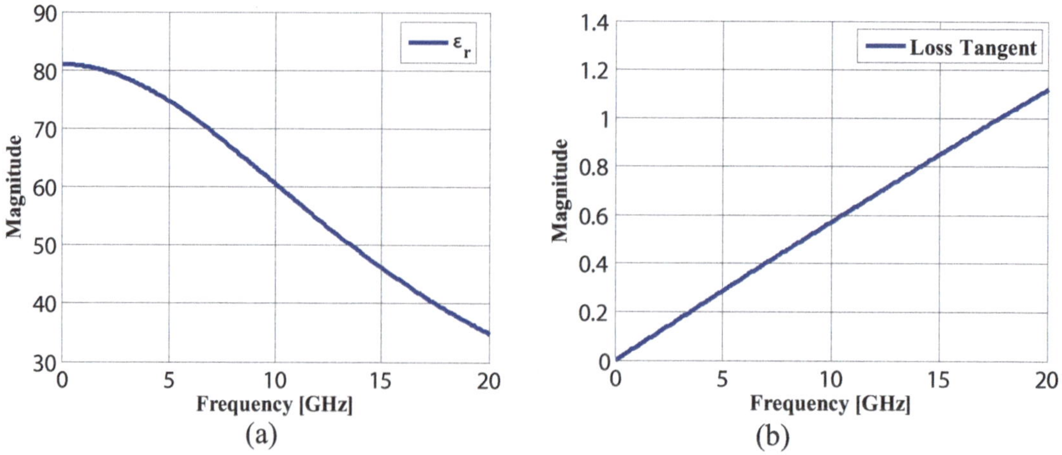

Figure 4.4: Water dispersive property versus frequency, (a) Relative permittivity, (b) Loss tangent.

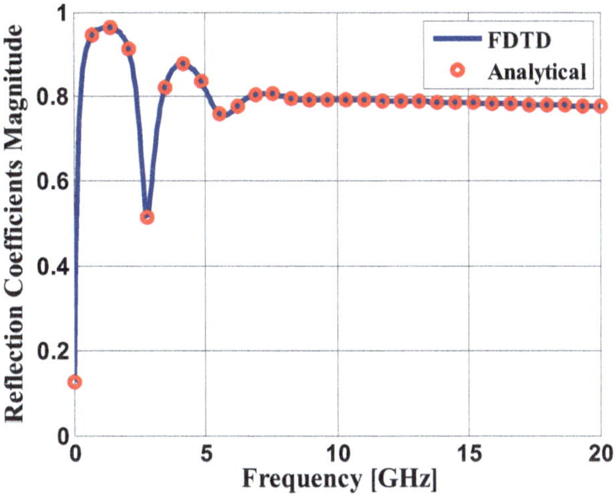

Figure 4.5: Reflection coefficient for infinite water slab of thickness 6 mm under normal incidence ($k_x = 0\text{m}^{-1}$) (from [44] © IEEE).

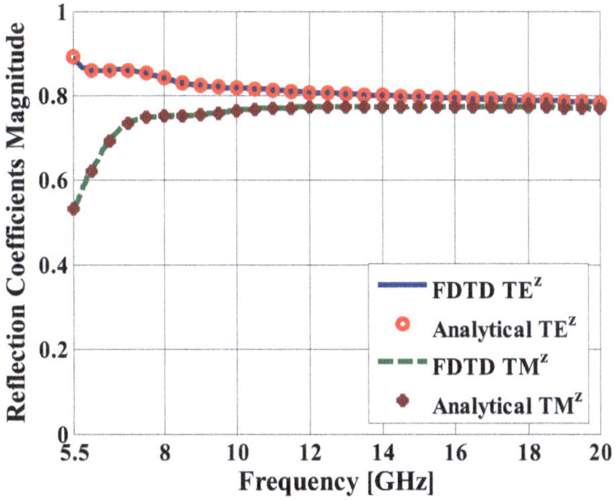

Figure 4.6: Reflection coefficients for infinite water slab of thickness 6 mm TM^z and TE^z oblique incidence ($k_x = 104.8\ \text{m}^{-1}$) (from [44] © IEEE).

4.4.2 NANOPLASMONIC SOLAR CELL STRUCTURE

The algorithm is then used to analyze a nanoplasmonic solar cell structure. The nanoparticles are used to increase the optical absorption within semiconductor solar cells, and hence enhance its performance [46]. The structure consists of cuboids elements of silver particles (dispersive media). The cuboids have length of 20 nm, width of 20 nm, and height of 10 nm. These cuboids are mounted over a SiO_2 (silicon dioxide) substrate of thickness 30 nm and ε_r = 3.9, and the structure has periodicity of 30 × 30 nm in both the x- and y-directions, as shown in Fig. 4.7. The permittivity of the

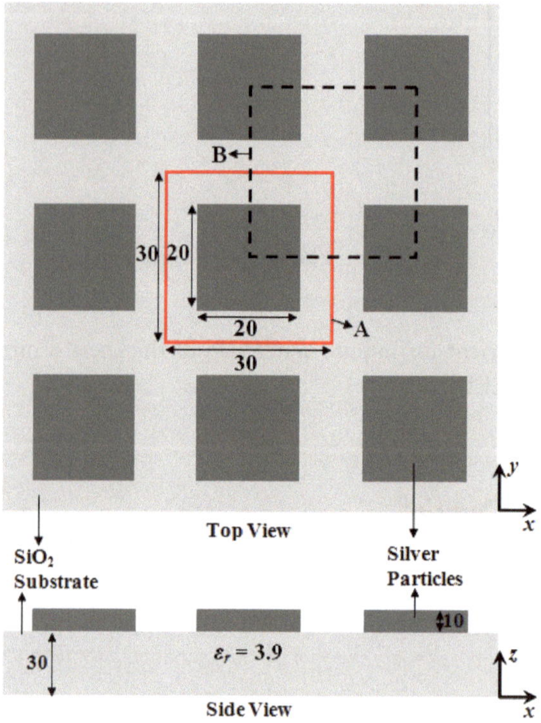

Figure 4.7: Geometry of the nanoplasmonic solar cell (all dimensions are in nm).

silver particles is described by a single-pole Lorentz medium using the parameters in [47, 48]. Using these parameters and Equation (4.10), the parameters for a two-term Debye relaxation model were derived (details are provided in Appendix A.4) as follows: $\varepsilon_{s1} = 4.8233\times10^{7}$, $\varepsilon_{s2} = -2.50721\times10^{5}$, $\varepsilon_\infty = 4.391$, $\tau_1 = 6.8479\times10^{-12}$ and $\tau_2 = 3.5597\times10^{-14}$. The dispersive properties of the silver versus frequency are shown in Fig. 4.8.

As shown in Fig. 4.7, the structure can be simulated using unit cell A or unit cell B. If the structure is simulated using unit cell A, the conventional PBC can be used since all the boundaries of the unit cell A are dielectric and there are no dispersive media on the boundaries (but the non-

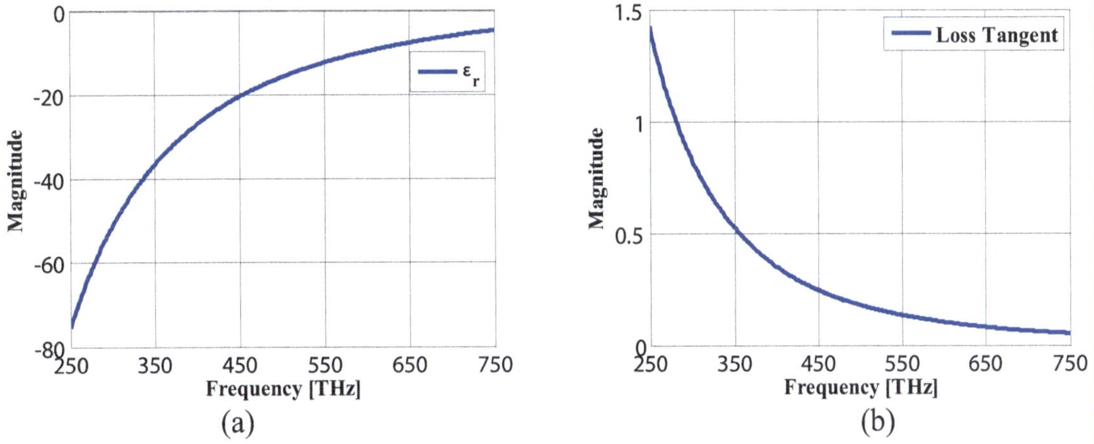

Figure 4.8: Silver dispersive property versus frequency, (a) Relative permittivity, (b) Loss tangent.

boundary field components still need to be handled using ADE technique). However, if the structure to be simulated uses unit cell B, the new DPBC must be used due to the presence of dispersive media on the boundaries. Figure 4.9 shows the simulation domains used in both cases. The structure is illuminated by a normally incident plane wave ($k_x = 0$ m^{-1}), using a cosine-modulated Gaussian pulse centered at 500 THz with a bandwidth of 500 THz. The structure is simulated using an FDTD grid cell size $\Delta x = \Delta y = \Delta z = 1.25$ nm, 50,000 time steps, and a 0.9 reduction factor of the CFL time step; the CPML is used for the absorbing boundaries at the top and the bottom of the computational domain.

The results of case 1 and case 2 are compared with results obtained from Ansoft HFSS in Fig. 4.10. Good agreement between the results generated using HFSS, FDTD case 1, and the new algorithm can be noticed. The computational time for FDTD case 1 is 120 minutes, and for case 2 is 121 minutes, while using HFSS for 40 frequency points requires 350 minutes. The good agreement between the results generated using FDTD case 1 (conventional constant horizontal wavenumber PBC) and results generated using FDTD case 2 (new DPBC) prove the validity of the new DPBC. In addition, it should be noticed that the presence of the nanoparticles enhances the absorption of the structure at the frequency range around 550 THz.

4.4.3 SANDWICHED COMPOSITE FSS

Next the algorithm is used to analyze a sandwiched composite-FSS structure. This composite material has been investigated for the potential applications as shielding materials to protect electronics system from electromagnetic pulses or electromagnetic interference. To enhance the shielding effectiveness, one possible solution is to introduce an additional layer or layers of FSS structures between

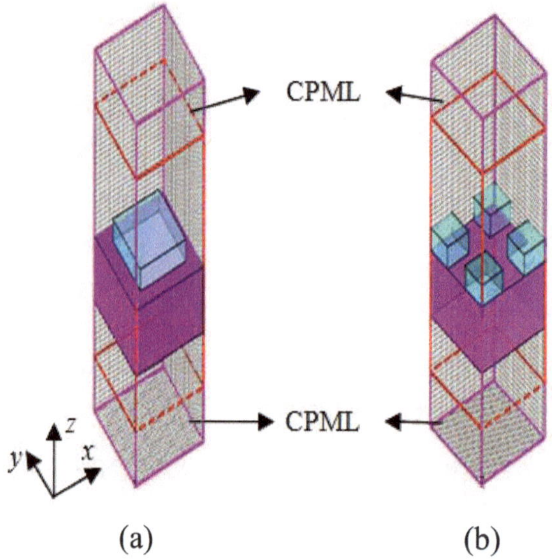

Figure 4.9: Simulation domain: (a) FDTD case 1 (unit cell A), (b) FDTD case 2 (unit cell B).

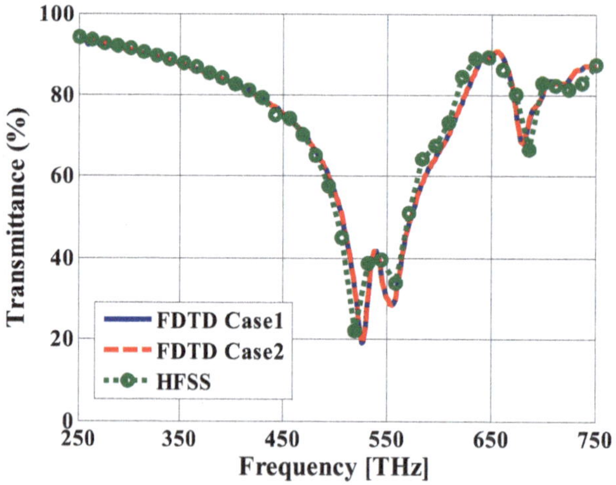

Figure 4.10: Transmittance (%) for the nanoplasmonic solar cells normal plane wave ($k_x = 0\,\mathrm{m}^{-1}$) case.

the interfaces of the composite material [49]. The sandwiched structure studied here is shown in Fig. 4.11.

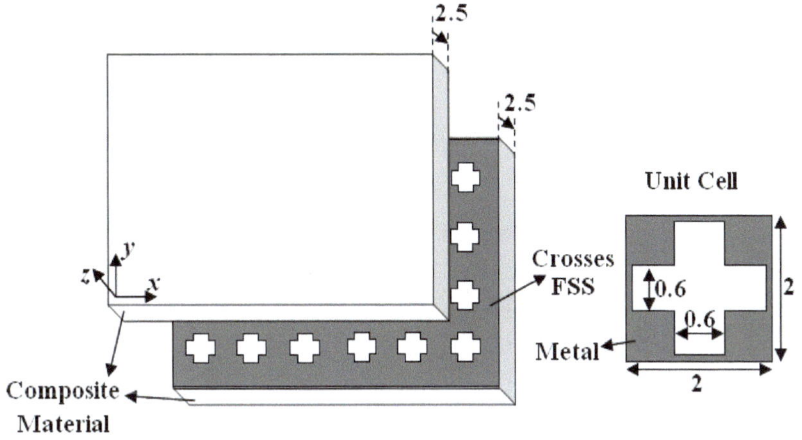

Figure 4.11: Geometry of the sandwiched composite-FSS structure (all dimensions are in mm) (from [44] © IEEE).

An infinite thin metal film is inserted between two composite material layers with a thickness of 2.5 mm each; the metal film has a periodic array of cross-shaped slots with a 2 mm periodicity in both the x- and y-directions. The parameters of the permittivity of the composite medium as given in [49] are: $\varepsilon_{s1} = 5.2$, $\varepsilon_{s2} = 3.7$, $\varepsilon_{\infty} = 3.7$, $\tau_1 = 5.27\times10^{-10}$, and $\tau_2 = 0$. The dispersive properties of the composite material versus frequency are shown in Fig. 4.12. The structure is simulated using an FDTD grid cell size $\Delta x = \Delta y = \Delta z = 0.1$ mm and a 0.9 reduction factor of the CFL time step are used; the CPML is used for the absorbing boundaries at the top and the bottom of the computational domain. The structure is first illuminated by a normally incident plane wave ($\theta = 0°$ and $\phi = 0°$), using a cosine-modulated Gaussian pulse centered at 5 GHz with 10 GHz bandwidth. Then the structure is illuminated by an obliquely incident plane wave ($\theta = 30°$ and $\phi = 60°$). To study the shielding enhancement provided by adding the FSS at the interface of the composite media, the transmission coefficient without the presence of the FSS is provided as a reference.

Figure 4.13 provides results for a normal incident plane wave ($\theta = 0°$ and $\phi = 0°$) exciting the composite-FSS structure. Good agreement between results obtained from the new FDTD/DPBC algorithm and HFSS can be noticed, which proves the efficiency and the validity of the new algorithm. The computational time for FDTD is 9.56 minutes while the computational time is using HFSS for 40 frequency points is 12.34 minutes. Figure 4.14 provides results for an oblique incident plane wave ($\theta = 30°$ and $\phi = 60°$) illuminating the composite-FSS structure. Good agreement between results obtained from the new FDTD/DPBC algorithm and HFSS can be noticed. The computational time for FDTD is equal to 13.6 minutes while the computational time using HFSS

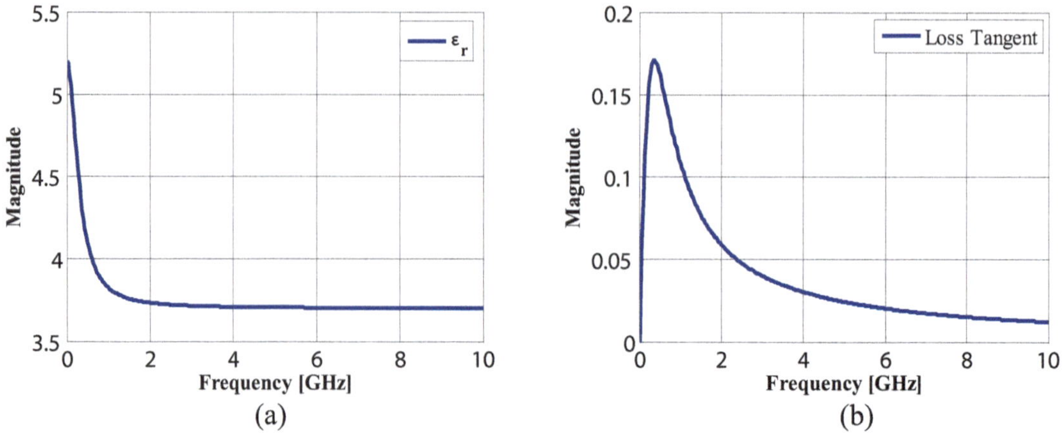

Figure 4.12: Composite material dispersive property versus frequency, (a) Relative permittivity, (b) Loss tangent.

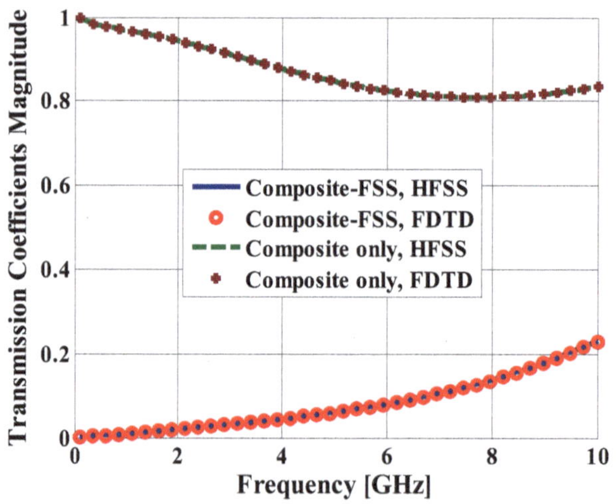

Figure 4.13: Transmission coefficient for sandwiched composite-FSS structure illuminated by a normally incident plane wave ($\theta = 0°$, $\phi = 0°$) (from [44] © IEEE).

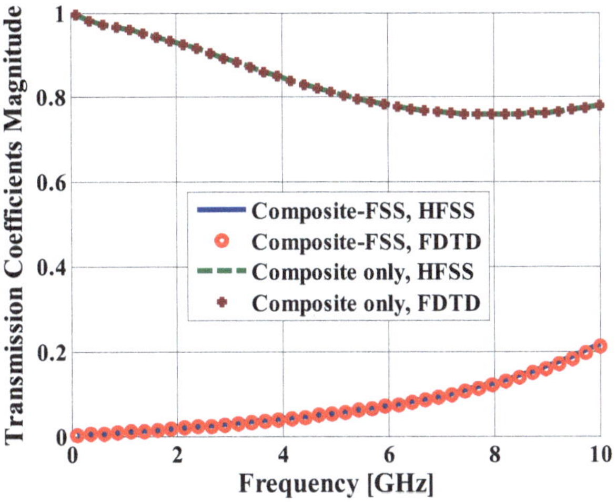

Figure 4.14: Transmission coefficient for sandwiched composite-FSS structure illuminated by an obliquely incident plane wave ($\theta = 30°$, $\phi = 60°$) (from [44] © IEEE).

for 40 frequency points is 14.8 minutes. It should also be noticed from Figs. 4.13 and 4.14 that the transmission coefficient is dramatically decreased due to the presence of the FSS, which enhance the shielding effectiveness.

4.5 SUMMARY

This chapter introduces a new FDTD/DPBC to analyze the scattering properties of periodic structures with dispersive media extended to the boundaries. The approach is developed based on both the constant horizontal wavenumber and the auxiliary differential equation techniques. The new procedure is simple to implement and efficient in terms of both computational time and memory usage. The algorithm is capable of calculating reflection and transmission coefficients for the cases of normal and oblique incidence and for both TE^z and TM^z polarizations. Numerical examples for potential applications such as dispersive slabs, nano-plasmonic structures, and sandwiched composite-FSS were provided. The results show good agreement with results from the analytical solution for a dispersive slab, and with the frequency-domain solutions for various dispersive periodic structures.

CHAPTER 5

Multilayered Periodic Structures

5.1 INTRODUCTION

Many periodic structures are built up of several layers, each layer being either a diffraction grating, periodic in one or two directions, or a homogenous dielectric slab which acts as a separator or support [50]. Two approaches can be employed to analyze multi-layer structures. One approach is to formulate and analyze a specific composite structure in its entirety [51]. This approach has serious practical limitations because the required amount of computation increases rapidly as the number of layers increases, and also because a complete new analysis is required every time a change is made in any layer. The other approach is to compute the generalized scattering matrix (GSM) [52, 53, 54] for each layer and then obtain the total GSM of the entire structure by simple matrix calculations. This approach is more flexible and applicable to practical problems where several layers may be cascaded in an arbitrary sequence. The cascading technique allows one to take advantage of different methods in computing the GSM for each layer of a multi-layer structure. In most of the previous work, the method of moments (MoM) and the finite element method (FEM) are used to compute the scattering parameters of each layer. In this chapter, the finite-difference time-domain (FDTD) with the constant horizontal wavenumber periodic boundary condition (PBC) approach described in Chapter 2 is used to compute the scattering parameters of each layer.

Usually, the GSM consists of scattering parameters of incident waves and their space harmonics, known as Floquet harmonics [55, 56]. In previous chapters all the simulations were full wave simulations for single layer periodic structures and that is why the Floquet harmonics were not mentioned. However, in multi-layer periodic structures, the Floquet harmonics are important due to the interactions between layers. Many parameters affect the behavior of the harmonics including the frequency range of interest, incident angle and polarization, periodicity and geometry of each layer, sequence of different layers, and separation between these layers. A complete Floquet harmonic analysis is presented in this chapter, where propagation and evanescent behaviors of harmonics are studied using the FDTD method. In addition, guidelines are provided to select proper higher order harmonics for certain separation sizes. It is worthwhile to point out that the FDTD algorithm used in this chapter is efficient for the harmonic analysis since the periodic boundary condition is handled by the constant horizontal wavenumber approach.

This chapter is organized as follows: In Section 5.2, different categories of multi-layer periodic structures are defined. In Section 5.3, the hybrid FDTD/GSM approach is described and the definition, computation, and conversion of scattering and transmission matrices are provided. In Section 5.4, a complete Floquet harmonic analysis of periodic structure is presented and the propagation and evanescent behaviors of the Floquet harmonics are studied. In addition, guidelines for harmonics selection are provided. Section 5.5 provides numerical examples to prove the validity of the hybrid FDTD/GSM approach. The algorithm is used to simulate various multi-layer periodic structures such as dipole and square patch FSS structures with different periodicities, and with normal and oblique incidences. The scattering properties of the entire multi-layer structures are calculated for both co- and cross-polarization components. In Section 5.6, a summary of the proposed algorithm is provided.

5.2 CATEGORIES OF MULTILAYERED PERIODIC STRUCTURES

Multilayered periodic structures could be categorized according to the periodicities of the layers or the separation between layers. As shown in Fig. 5.1, three categories exist according to the periodicity of different layers: in the first category, all the layers have the same periodicities (which will be referred to as the 1:1 case); in the second category, the periodicities of one layer is integer multiples of another layer (which will be referred to as the n:1 case); in the third category, the periodicities of the layers are not integer multiples of each other (which will be referred to as the n:m case). As shown in Fig. 5.1, in the first category (1:1), the two layers have the same periodicity of 15 mm × 15 mm in this specific case. In the second category (n:1), the first layer has a periodicity of 7.5 mm × 7.5 mm, and the second layer has a periodicity of 15 mm × 15 mm (4 unit cells : 1 unit cell). In the third category (n:m), the first layer has a periodicity of 10 mm × 10 mm, and the second layer has a periodicity of 15 mm × 15 mm (9 unit cells : 4 unit cells).

As shown in Fig. 5.2, two categories exist according to the separation between layers. In the first category, the separation between layers is large enough to neglect the effects of the higher order harmonics (which will be referred to as the large gap case). In the second category, the separation between layers is small so that the effects of the higher order harmonics cannot be neglected (which will be referred to as the small gap case).

5.3 HYBRID FDTD/GSM METHOD

In this section, the hybrid FDTD/GSM approach is described. The definition, computation, and conversion of scattering and transmission matrices are provided.

5.3.1 PROCEDURE OF HYBRID FDTD/GSM METHOD

As described here for multi-layer periodic structures, the GSM technique can take into account propagating and non-propagating modes and interactions between them (including cross-polarization effects). It describes the reflection and transmission properties of each layer by a scattering matrix and uses a cascading process to obtain a scattering matrix for the overall structure. The modes are the Floquet spatial harmonics of a plane-wave incident on a structure with specified periodicity. Each element in the scattering matrix is either a reflection or a transmission coefficient, which provides

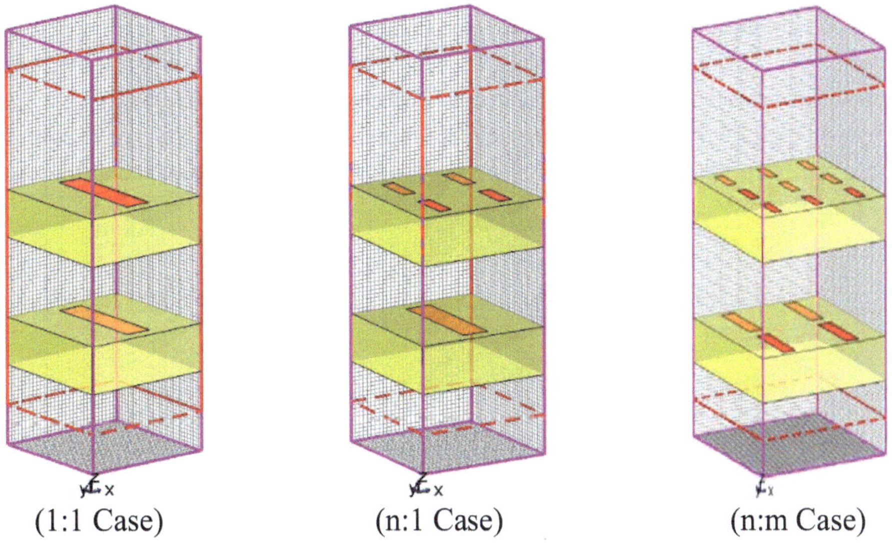

Figure 5.1: Three categories of multi-layer periodic structures according to the periodicity.

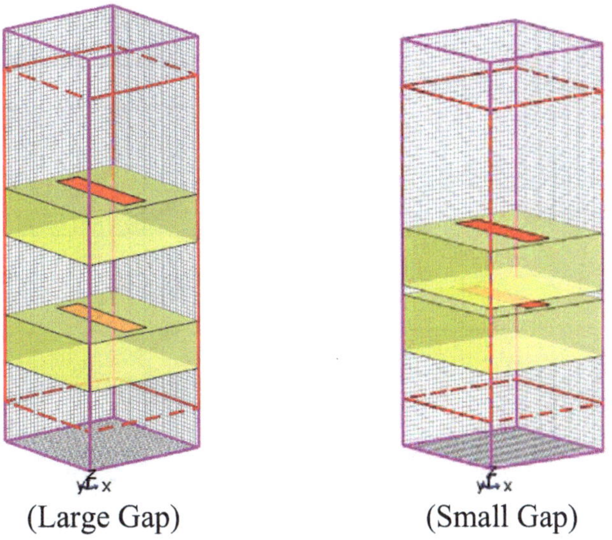

Figure 5.2: Two categories of multi-layer periodic structures according to the separation.

the linear relationship between a scattered harmonic and one of the incident harmonics that excites it. The scattering matrix for a single layer can be transformed into a transmission matrix, and the cascading procedure is applied to the single-layer transmission matrices to produce a transmission matrix for the overall structure. This matrix can then be transformed to produce a scattering matrix for the overall structure. In principle, the solution accuracy can be improved by using a large matrix for each layer to include more Floquet harmonics. In practice, the objective is to choose the matrix size large enough for good accuracy but small enough to keep the expenditure of computing resources within acceptable limits.

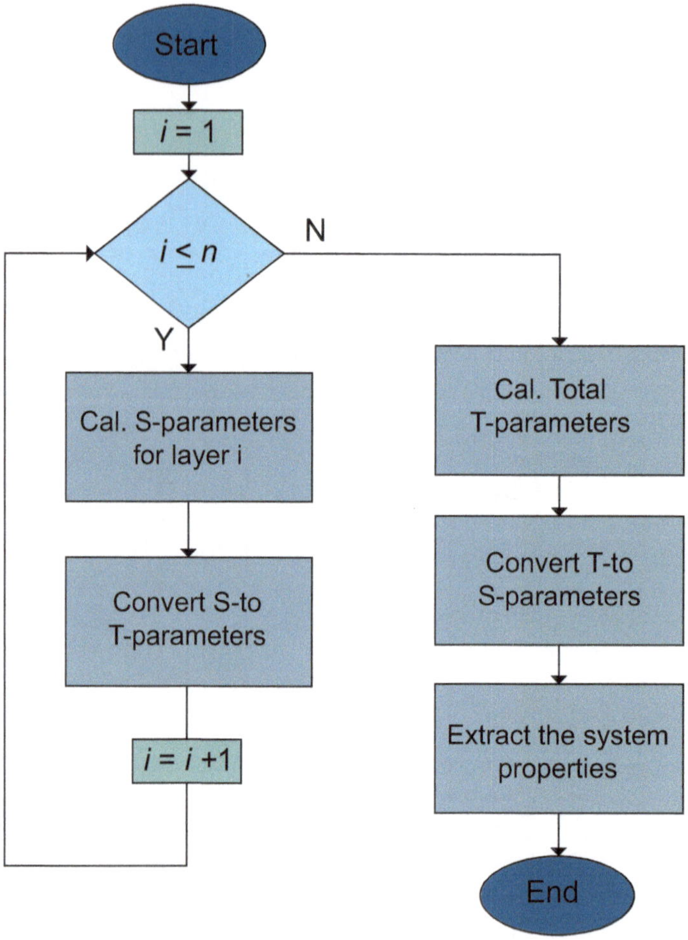

Figure 5.3: Flow chart of the hybrid FDTD/GSM algorithm.

As shown in Fig. 5.3, the proposed algorithm can be summarized as follows:

1) Using the constant horizontal wavenumber FDTD/PBC, the scattering parameters of the first layer are calculated and the scattering matrix is constructed.

2) The scattering matrix of the first layer is transformed to a transmission matrix.

3) Step 1 and 2 are repeated for all the layers.

4) The total transmission matrix is calculated using matrix multiplication for all the transmission matrices.

5) The total transmission matrix is transformed to a scattering matrix and all the scattering parameters are extracted from it.

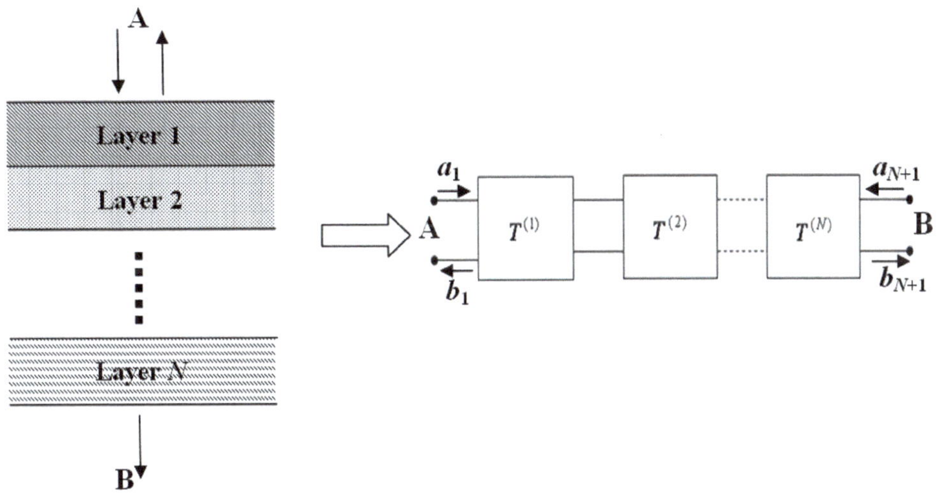

Figure 5.4: Multilayered periodic medium and its equivalent transmission matrices.

For the layered medium shown in Fig. 5.4, the total composite transmission matrix is given by

$$T_{total} = \underset{\sim}{T}^{(N)} \cdots \underset{\sim}{T}^{(2)} \underset{\sim}{T}^{(1)}, \tag{5.1}$$

where the transmission and scattering matrices are defined as

$$\begin{bmatrix} b_1 \\ a_1 \end{bmatrix} = \begin{bmatrix} T_{11} & T_{12} \\ T_{21} & T_{22} \end{bmatrix} \begin{bmatrix} a_2 \\ b_2 \end{bmatrix}, \quad \begin{bmatrix} b_1 \\ b_2 \end{bmatrix} = \begin{bmatrix} S_{11} & S_{12} \\ S_{21} & S_{22} \end{bmatrix} \begin{bmatrix} a_1 \\ a_2 \end{bmatrix}. \tag{5.2}$$

The transformations between the $[S]$ and $[T]$ matrices are given by

$$[T] = \begin{bmatrix} S_{12} - S_{11}S_{21}^{-1}S_{22} & S_{11}S_{21}^{-1} \\ -S_{21}^{-1}S_{22} & S_{21}^{-1} \end{bmatrix}, \tag{5.3a}$$

$$[S] = \begin{bmatrix} T_{12}T_{22}^{-1} & T_{11} - T_{12}T_{22}^{-1}T_{21} \\ T_{22}^{-1} & -T_{22}^{-1}T_{21} \end{bmatrix}. \tag{5.3b}$$

When cross-polarization components or higher harmonics are included, T_{ij} and S_{ij} of (5.3) become sub-matrices, and the variables a_j and b_j become vectors. Equation (5.3) can be easily proved for the general case using matrix partitioning as shown in Appendix B.1.

5.3.2 CALCULATING SCATTERING PARAMETERS USING FDTD/PBC

In this section, the scattering parameters of a single layer periodic structure are calculated using the constant horizontal wavenumber FDTD/PBC technique described in Chapter 2. We consider a case of a single layer periodic structure, where the layer is periodic in both the x- and y-directions and is illuminated by a plane wave with general oblique incidence as shown in Fig. 5.5. Using the constant horizontal wavenumber FDTD/PBC technique, only one unit cell is simulated to get the scattering parameters of the entire layer. Let's start with a simple case where only the co- and cross-polarization components of the dominant mode (without any higher order Floquet harmonics) are calculated.

As shown in Fig. 5.6, a_1, b_1, a_3, and b_3 are related to the co-polarized electric field components of the dominant mode, while a_2, b_2, a_4, and b_4 are related to the cross-polarized electric field components of the dominant mode. Four different field components exist, so the scattering matrix will be of the size 4×4. The S-parameters are calculated as

$$S_{11} = \frac{E_r^{Co-pol}}{E_{i(top)}^{Co-pol}}, \quad S_{21} = \frac{E_r^{X-pol}}{E_{i(top)}^{Co-pol}}, \quad S_{31} = \frac{E_t^{Co-pol}}{E_{i(top)}^{Co-pol}}, \quad S_{41} = \frac{E_t^{X-pol}}{E_{i(top)}^{Co-pol}}, \tag{5.4a}$$

$$S_{12} = \frac{E_r^{Co-pol}}{E_{i(top)}^{X-pol}}, \quad S_{22} = \frac{E_r^{X-pol}}{E_{i(top)}^{X-pol}}, \quad S_{32} = \frac{E_t^{Co-pol}}{E_{i(top)}^{X-pol}}, \quad S_{42} = \frac{E_t^{X-pol}}{E_{i(top)}^{X-pol}}. \tag{5.4b}$$

For the rest of the S-parameters, the plane wave excitation is placed below the layer for an unsymmetrical layer,

$$S_{13} = \frac{E_t^{Co-pol}}{E_{i(bottom)}^{Co-pol}}, \quad S_{23} = \frac{E_t^{X-pol}}{E_{i(bottom)}^{Co-pol}}, \quad S_{33} = \frac{E_r^{Co-pol}}{E_{i(bottom)}^{Co-pol}}, \quad S_{43} = \frac{E_r^{X-pol}}{E_{i(bottom)}^{Co-pol}}, \tag{5.4c}$$

$$S_{14} = \frac{E_t^{Co-pol}}{E_{i(bottom)}^{X-pol}}, \quad S_{24} = \frac{E_t^{X-pol}}{E_{i(bottom)}^{X-pol}}, \quad S_{34} = \frac{E_r^{Co-pol}}{E_{i(bottom)}^{X-pol}}, \quad S_{44} = \frac{E_r^{X-pol}}{E_{i(bottom)}^{X-pol}}, \tag{5.4d}$$

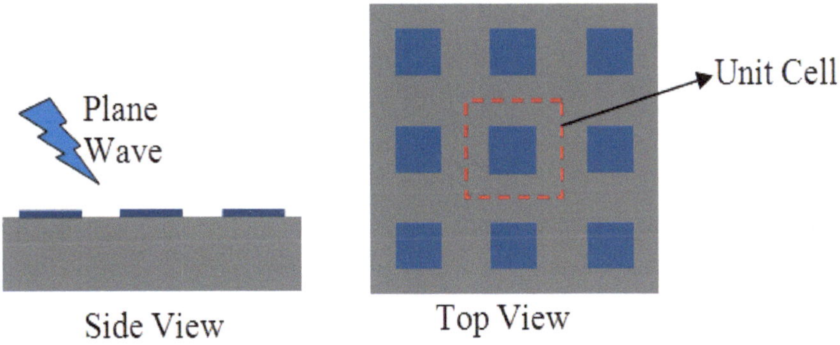

Figure 5.5: Geometry of single layer periodic structure.

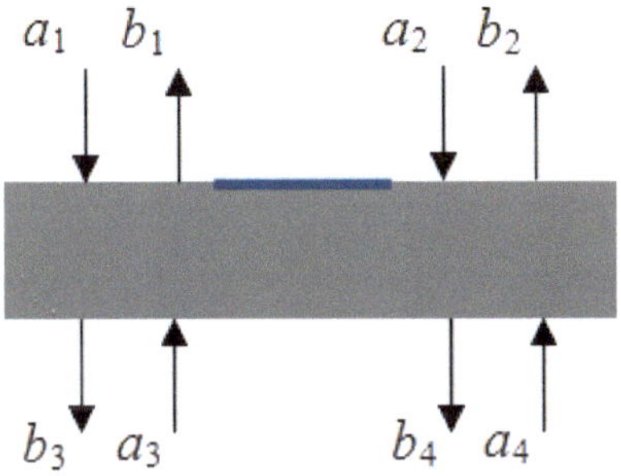

Figure 5.6: Reflected and transmitted electric fields for co- and cross-polarized components of the dominant mode.

where $E_{t/r/i}^{Co/x-pol}$ are the complex amplitudes of the frequency-domain electric fields [57, 58], which can be obtained from the time-domain electric field by using the discrete Fourier Transform (DFT). For the TE^z plane wave, the co- and cross-polarized components can be stated as follows [6]:

$$E^{Co-pol} = E_y \frac{k_x}{\sqrt{k_x^2 + k_y^2}} - E_x \frac{k_y}{\sqrt{k_x^2 + k_y^2}}, \tag{5.4e}$$

$$E^{X-pol} = E_x \frac{k_x}{\sqrt{k_x^2 + k_y^2}} + E_y \frac{k_y}{\sqrt{k_x^2 + k_y^2}}.$$

The S-parameters for other layers can be calculated similarly and transformed to T-parameters as shown in (5.3). Symmetry can be used to reduce the calculation of S-parameters (only the excitation above the layer will be enough). As for dielectric layers or air gaps, the homogeneity of these layers decreases the S-parameter calculation simulation time (it can be also calculated analytically).

Similar to the cross-polarization analysis, any number of higher-order Floquet harmonics can be added, and the S-parameters due to these higher harmonics can be calculated. The decomposition of electric field periodic in two dimensions is of the form:

$$\vec{E}(x,y,z)=\sum_{n}\sum_{m}\vec{A}_{m,n}e^{j(k_x^{m,n}x+k_y^{m,n}y+k_z^{m,n}z)}, \tag{5.5}$$

where the $\vec{A}_{m,n}$ are the vector coefficients of the decomposition, and $k_x^{m,n}$ and $k_y^{m,n}$ are the wavenumbers of the Floquet modes determined by the cell dimensions of the periodic structure. The wavenumber of the incident field are defined as follows:

$$k_x^{m,n}=k\sin\theta\cos\phi+\frac{2\pi m}{P_x}, \tag{5.6a}$$

$$k_y^{m,n}=k\sin\theta\cos\phi+\frac{2\pi n}{P_y\sin\alpha}-\frac{2\pi m}{P_x\tan\alpha}, \tag{5.6b}$$

where P_x , P_y , and α describe the geometry of the unit cell as shown in Fig. 5.7, and α is the skew

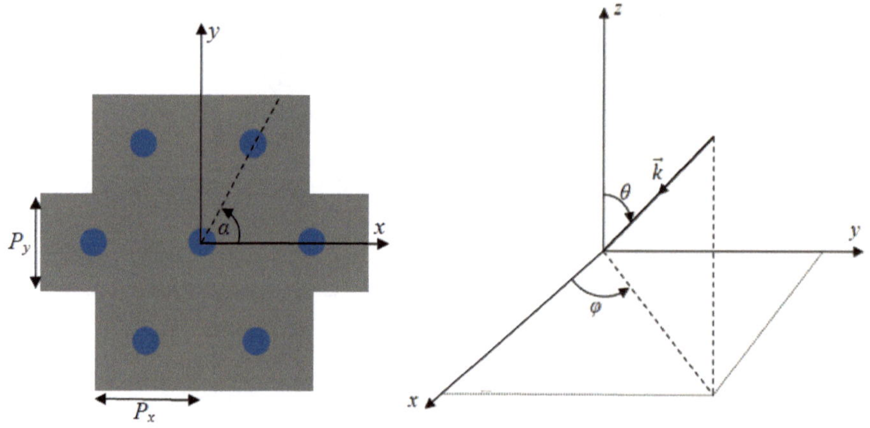

(a) (b)

Figure 5.7: (a) Two dimensional periodic scatterer, (b) General incident plane wave.

angle of the grid. In this chapter, this angle α will be taken as 90° (axial case), so the Equation (5.6b) will be rewritten as

$$k_y^{m,n}=k\sin\theta\cos\phi+\frac{2\pi n}{P_y}, \tag{5.6c}$$

and the wavenumber in the z-direction is defined as:

$$k_z^{m,n}=\sqrt{k^2-(k_x^{m,n})^2-(k_y^{m,n})^2}, \tag{5.7}$$

where $k_z^{m,n}$ is real for propagating modes and imaginary for non-propagating modes. Each element of a scattering matrix in (5.2) is a scattering parameter, either a reflection coefficient or a transmission coefficient, that gives the linear relationship between the amplitude of a scattered harmonic ($\vec{A}_{m,n}$) and one of the incident harmonics that excites it ($\vec{A}_{i,j}$) [50].

To illustrate the above procedure, analyzing a periodic layer while taking into account only two modes (the dominant mode and the first harmonic) is considered. The same procedure is applied for the co- and cross-polarized components. Thus, the S-parameters are calculated as follows:

$$S_{11} = \frac{E_r^{Dom}}{E_{i(top)}^{Dom}}, \quad S_{21} = \frac{E_r^{Harm1}}{E_{i(top)}^{Dom}}, \quad S_{31} = \frac{E_t^{Dom}}{E_{i(top)}^{Dom}}, \quad S_{41} = \frac{E_t^{Harm1}}{E_{i(top)}^{Dom}}, \tag{5.8a}$$

$$S_{12} = \frac{E_r^{Dom}}{E_{i(top)}^{Harm1}}, \quad S_{22} = \frac{E_r^{Harm1}}{E_{i(top)}^{Harm1}}, \quad S_{32} = \frac{E_t^{Dom}}{E_{i(top)}^{Harm1}}, \quad S_{42} = \frac{E_t^{Harm1}}{E_{i(top)}^{Harm1}}. \tag{5.8b}$$

For the rest of the S parameters, the plane wave excitation is placed below the layer (for an unsymmetrical layer):

$$S_{13} = \frac{E_t^{Dom}}{E_{i(bottom)}^{Dom}}, \quad S_{23} = \frac{E_t^{Harm1}}{E_{i(bottom)}^{Dom}}, \quad S_{33} = \frac{E_r^{Dom}}{E_{i(bottom)}^{Dom}}, \quad S_{43} = \frac{E_r^{Harm1}}{E_{i(bottom)}^{Dom}}, \tag{5.8c}$$

$$S_{14} = \frac{E_t^{Dom}}{E_{i(bottom)}^{Harm1}}, \quad S_{24} = \frac{E_t^{Harm1}}{E_{i(bottom)}^{Harm1}}, \quad S_{34} = \frac{E_r^{Dom}}{E_{i(bottom)}^{Harm1}}, \quad S_{44} = \frac{E_r^{Harm1}}{E_{i(bottom)}^{Harm1}}, \tag{5.8d}$$

where $E_{t/r/i}^{Dom/Harm1}$ are the complex amplitudes of the frequency-domain electric fields for both the dominant mode and first harmonic (incident or reflected or transmitted). Calculating these complex amplitudes is described in detail in the next section.

5.4 FDTD/PBC FLOQUET HARMONIC ANALYSIS OF PERIODIC STRUCTURES

In this section, a procedure is developed to extract all the harmonics from the FDTD/PBC simulation and study their frequency behavior. In addition, another procedure is developed based on the geometric properties of the multi-layer periodic structure and the frequency-domain harmonic behavior to determine the proper gap size after which higher harmonic effects can be neglected. The latter procedure is also used as a guideline to select the proper harmonics to be considered in the analysis for a certain gap size.

5.4.1 EVANESCENT AND PROPAGATION HARMONICS IN PERIODIC STRUCTURES

The presence of periodicity in the scatterer can lead to the appearance of far-field transmission and reflection at additional angles, often referred to as Floquet harmonics [18]. In this book the periodicity is in the x- and y-directions, and the generated harmonics will have wavenumbers as follows:

$$k_x^{m,n} = k_x^i + \frac{2\pi m}{P_x}, \quad k_y^{m,n} = k_y^i + \frac{2\pi n}{P_y}, \tag{5.9}$$

where m and n are the harmonic indices in the x- and y-directions, respectively.

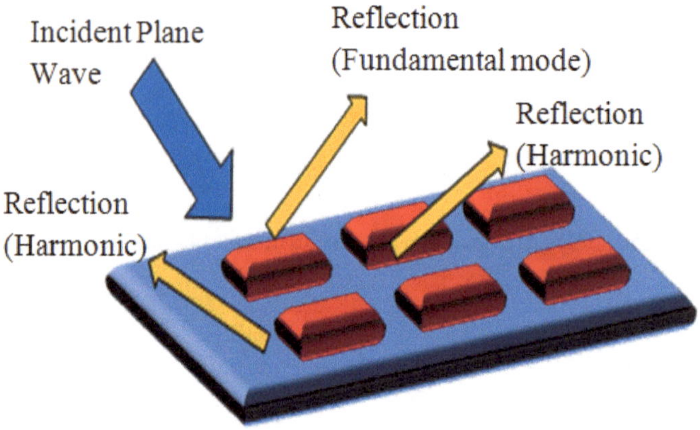

Figure 5.8: Incident plane wave and the reflected harmonics.

In this analysis, the harmonics are named using the following convention:

$$M_{m,n} \rightarrow m = 0, \pm 1, \pm 2 \cdots \text{ and } n = 0, \pm 1, \pm 2 \cdots . \tag{5.10}$$

For example, the basic (dominant) mode and two different harmonics are as follows:

$$\begin{aligned} m = 0, n = 0 &\rightarrow M_{0,0}\left(k_x^{0,0} = k_x^i, k_y^{0,0} = k_y^i\right), \\ m = 1, n = -1 &\rightarrow M_{1,-1}\left(k_x^{1,-1} = k_x^i + \frac{2\pi}{P_x}, k_y^{1,-1} = k_y^i - \frac{2\pi}{P_y}\right), \\ m = -2, n = 4 &\rightarrow M_{-2,4}\left(k_x^{-2,4} = k_x^i - \frac{4\pi}{P_x}, k_y^{-2,4} = k_y^i + \frac{8\pi}{P_y}\right). \end{aligned} \tag{5.11}$$

These harmonics have cut-off frequencies, after which the harmonics start to propagate and are no longer evanescent harmonics. To determine the cut-off frequencies of different harmonics, consider the case of normal incidence where $k_x^i = k_y^i = 0$, and consider a periodic structure with 15 mm × 15 mm. Using this information, the cut-off frequencies of the first five modes can be calculated as follows:

$$\begin{aligned}
&m = 0, n = 0 \to M_{0,0} \to k_x^{0,0} = 0, k_y^{0,0} = 0,\\
&m = 0, n = 1 \to M_{0,1} \to k_x^{0,1} = 0, k_y^{0,1} = 0 + \frac{2\pi}{15 \times 10^{-3}} = 418.879,\\
&m = 0, n = -1 \to M_{0,-1} \to k_x^{0,-1} = 0, k_y^{0,-1} = 0 - \frac{2\pi}{15 \times 10^{-3}} = -418.879,\\
&m = 1, n = 0 \to M_{1,0} \to k_x^{1,0} = 0 + \frac{2\pi}{15 \times 10^{-3}} = 418.879, k_y^{1,0} = 0,\\
&m = -1, n = 0 \to M_{-1,0} \to k_x^{-1,0} = 0 - \frac{2\pi}{15 \times 10^{-3}} = -418.879, k_y^{-1,0} = 0.
\end{aligned}$$

At $k^2 = (k_x^{m,n})^2 + (k_y^{m,n})^2$, the cut-off frequency occurs, which can be calculated as follows for different modes:
$k = \frac{2\pi f}{c}$ for free space where c is the speed of light in free space.

$$\begin{aligned}
&m = 0, n = 0 \to k = \sqrt{(0)^2 + (0)^2} = 0 \to f_{\textit{cut-off}}^{0,0} = 0\text{GHz},\\
&m = 0, n = 1 \to k = \sqrt{(0)^2 + (418.879)^2} = 418.879 \to f_{\textit{cut-off}}^{0,1} \simeq 20\text{GHz},\\
&m = 0, n = -1 \to k = \sqrt{(0)^2 + (-418.879)^2} = 418.879 \to f_{\textit{cut-off}}^{0,-1} \simeq 20\text{GHz},\\
&m = 1, n = 0 \to k = \sqrt{(418.879)^2 + (0)^2} = 418.879 \to f_{\textit{cut-off}}^{1,0} \simeq 20\text{GHz},\\
&m = -1, n = 0 \to k = \sqrt{(-418.879)^2 + (0)^2} = 418.879 \to f_{\textit{cut-off}}^{-1,0} \simeq 20\text{GHz}.
\end{aligned}$$

To study the frequency behavior of the electric fields for these harmonics, the same previous assumptions (periodicity of 15 mm × 15 mm and normal incidence) will be considered, and the electric field of any mode can be generally written as (assume the y-component):

$$\vec{E} = E_0 e^{-j(k_x^{m,n}x + k_y^{m,n}y + k_z^{m,n}z)}\hat{a}_y. \tag{5.12}$$

Assuming that the magnitude of the incident electric field of each mode is unity and observing the electric field magnitude at a distance of 15 mm from the excitation plane, the attenuation of the magnitude of the electric field versus frequency is plotted in Fig. 5.9. It can be noticed from the figure that the cut-off frequency for the ($M_{1,0}$) harmonic is 20 GHz, as it was calculated analytically. Also, after that cut-off frequency, the harmonic starts to propagate. In addition, it can be noticed that the ($M_{1,1}$) harmonic cut-off frequency is almost 28.3 GHz. Moreover, it can be noticed that the effect of the harmonics increases for frequencies near the cut-off frequency. For the practical case of periodic structure, the magnitude of the harmonics (E_0) in (5.12) will not be unity, but it will be of a certain value depending on the angle of incidence and the geometry of the periodic structure. To calculate the actual magnitude of different harmonics, the expression for the magnitude of a harmonic related to a total field can be stated as follows:

$$E^{m,n}(\omega) = \frac{1}{P_x P_y}\int_0^{P_y}\int_0^{P_x} E(\omega, x, y)e^{jk_x^{m,n}x}e^{jk_y^{m,n}y}dxdy, \tag{5.13}$$

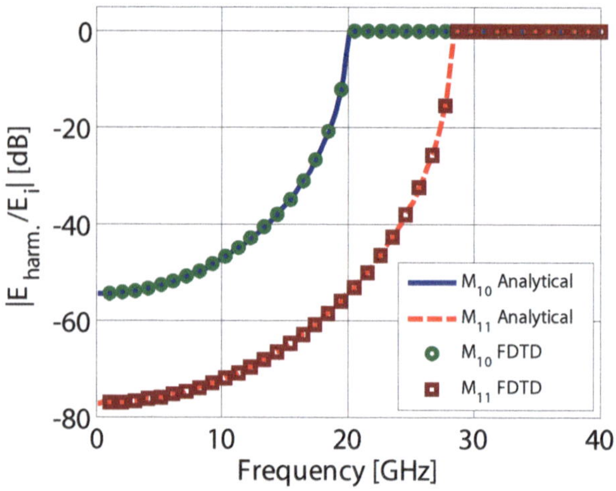

Figure 5.9: The magnitude of the electric field at 15 mm from the excitation plan for ($M_{1,0}$) and ($M_{1,1}$) harmonics.

where $k_x^{m,n}$, $k_y^{m,n}$ are given by Equation (5.9), $E(\omega, x, y)$ is the total frequency-domain field, and x, y are the position of this electric field. Equation (5.13) can be re-written in the discretized form as

$$E^{m,n}(\omega) = \frac{1}{N_x N_y} \sum_{u=0}^{N_x} \sum_{v=0}^{N_y} E(\omega, u\Delta x, v\Delta y) e^{jk_x^{m,n}(u\Delta x)} e^{jk_y^{m,n}(v\Delta y)}, \tag{5.14}$$

where N_x and N_y are the total number of cells in the x- and y-directions, respectively, and Δx and Δy are the cell size in the x- and y-directions, respectively. Moreover, $P_x = N_x \times \Delta x$ and $P_y = N_y \times \Delta y$.

The electric field $E(\omega, u\Delta x, v\Delta y)$ is calculated after the transformation of the time-domain electric field at each cell into the frequency-domain using the DFT. Traditionally, this process requires saving all the time-domain components of electric fields at each cell. For instance, if the simulation is performed for 30×30 cells and 2,500 time steps, for every time step, at least two matrices (E_x, E_y) of the size 30×30 have to be stored. These matrices are then transformed to the frequency-domain, and the magnitude of different harmonics can be calculated using (5.14), which requires huge memory usage. However, if the constant horizontal wavenumber approach is used, then $k_x^{m,n}$, $k_y^{m,n}$ are constant and (5.14) can be directly transformed to the time-domain as

$$E^{m,n}(t) = \frac{1}{N_x N_y} \sum_{u=0}^{N_x} \sum_{v=0}^{N_y} E(t, u\Delta x, v\Delta y) e^{jk_x^{m,n}(u\Delta x)} e^{jk_y^{m,n}(v\Delta y)}. \tag{5.15}$$

Using (5.15), the time-domain magnitude of each harmonic can be easily calculated in the FDTD/PBC simulation. Then this time-domain data is transformed to the frequency-domain using the DFT, which does not require any extra memory compared to the conventional FDTD technique, due to the fact that the fields are captured in the code using (5.15). This feature of the constant horizontal wavenumber FDTD/PBC approach is considered an important advantage because of the reduction in memory usage.

FDTD Harmonic Analysis Procedure:

1) Use constant horizontal wavenumber approach to calculate $E(t, u\Delta x, v\Delta y)$.
2) Use (5.15) to calculate the time-domain magnitude of different harmonics in the FDTD/PBC simulation.
3) Repeat steps 1 and 2 until all time steps in the FDTD simulation are completed.
4) Use the DFT to calculate the frequency-domain magnitude of different harmonics.

The above procedure can be used with any periodic structure to completely study the effect of different harmonics on the cascading configuration.

5.4.2 GUIDELINE FOR HARMONIC SELECTION

In this section a procedure for determining the proper gap size in order to neglect the higher harmonics effects is described. The procedure can also be used to determine which harmonics to be considered for specific gap size.

FDTD Gap Size Determination Procedure:

1) Specify the periodicity, the order, and the geometry of each layer: The periodicity and geometry of the layer are important to determine the cut-off frequencies and magnitudes of different harmonics. The layers order determines which of the reflected or transmitted harmonics are to be considered.
2) Specify the frequency range of interest. The frequency range of interest is important to determine whether the harmonics are propagating or evanescent harmonics in this frequency range.
3) Specify the incident wave parameters (k_x^i and k_y^i). Use k_x^i and k_y^i to determine the cut-off frequencies of different harmonics. Any propagating harmonics in the frequency range of interest should be considered whatever the gap size is.
4) Use the harmonic analysis procedure to determine the magnitudes of the evanescent harmonics: Calculate $k_z^{m,n}$ and use it together with the harmonic magnitudes to study the decaying behavior of the evanescent harmonics with distance.
5) The gap size for neglecting specific harmonic effect is calculated as the distance after which all evanescent harmonics magnitudes decay below -40 dB compared to the excitation level of

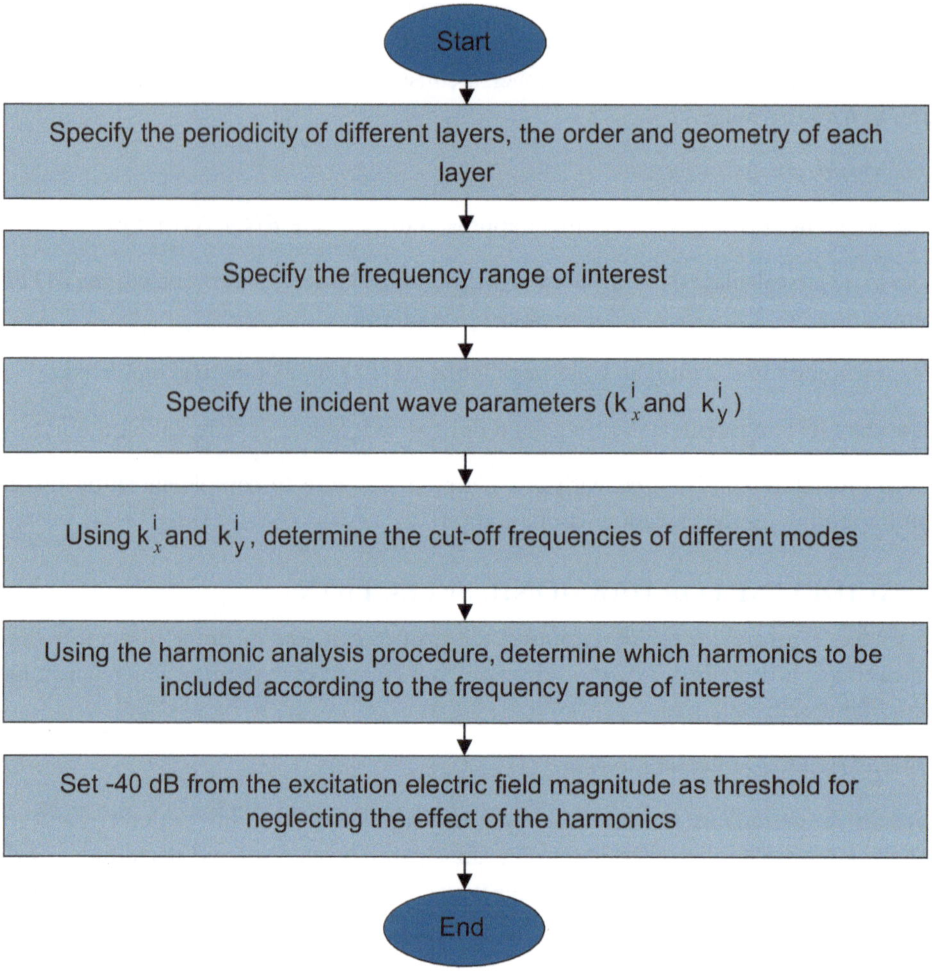

Figure 5.10: The flow chart of gap size determination procedure.

the corresponding field component magnitude. The −40 dB threshold was concluded from different test cases for error less than 5%. Other accuracy can be achieved by changing the threshold value.

If the gap size is less than that determined by this procedure, all the evanescent harmonics that have magnitudes larger than the threshold level should be included in the cascading process for accurate results.

5.5 NUMERICAL RESULTS

In this section, numerical examples are provided to prove the validity of the proposed algorithm. A test plan is summarized in Table 5.1, which covers different multi-layer categories described in Section 5.2 with different types of plane wave incidence (normal and oblique). In all the test cases, the results of the cascading technique are compared with the FDTD simulation of the entire structure. The FDTD code was developed in MATLAB and run on a computer with an Intel Core 2 CPU 6700, 2.66 GHz with 2 GB RAM.

Table 5.1: Test plan of multi-layer periodic structure analysis code

Test Case Number	Periodicity / Gap size	Incidence	Cross-polarization	Higher harmonics
1	Dielectric slab	Normal / Oblique	No	No
2	1:1 / Large	Normal	No	No
3	1:1 / Small	Normal	No	Yes
4	1:1 / Large	Oblique	Yes	No
5	1:1 / Small	Oblique	Yes	Yes
6	n:m / Large	Normal	No	No
7	n:m / Small	Normal	No	Yes
8	n:m / Large	Oblique	Yes	No

5.5.1 TEST CASE 1 (INFINITE DIELECTRIC SLAB)

Due to the homogeneity of the dielectric slab, it is considered a good verification case. In addition, the results can be compared with the analytical solution. The code is used to analyze an infinite dielectric slab with thickness h = 9.375 mm and relative permittivity ε_r = 2.56. The slab is illuminated by TM^z and TE^z plane waves, respectively. The plane wave is incident normally ($k_x = k_y$ = 0 m^{-1}) and obliquely (k_x = 104.8 m^{-1}, k_y = 0 m^{-1} for min frequency of 5 GHz).

In this test case, a dielectric slab with half the thickness of the original slab was simulated, and the cascading technique was used to simulate the original dielectric slab. As shown in Fig. 5.11, (a) the dielectric slab is analyzed analytically with different excitation polarizations and angles of incidence; (b) half the dielectric slab is analyzed using the FDTD/PBC technique and the scattering parameters are extracted as previously described; and (c) the cascading technique is used to get the scattering parameters of the whole dielectric slab from the scattering parameters of half of the original slab. The slab is excited using a cosine-modulated Gaussian pulse centered at 10 GHz with a 20 GHz bandwidth. The FDTD grid cell size is $\Delta x = \Delta y = \Delta z = 0.3125$ mm, and the slab is represented by 5x5 cells. In the FDTD code, 2,500 time steps and a 0.9 reduction factor of CFL time step are used. The CPML is used as the absorbing boundaries at the top and the bottom of the computational domain. The results are compared with analytical results in Figs. 5.12, 5.13, and 5.14. It should also be noticed that due to the homogeneity of the dielectric slab, the harmonics effects

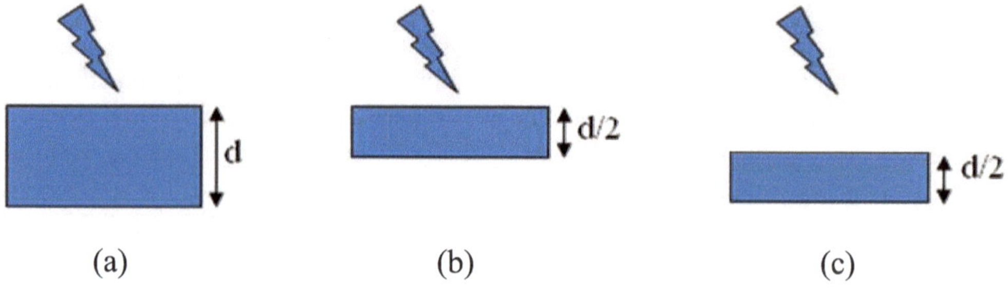

Figure 5.11: Dielectric slab simulation using cascading technique.

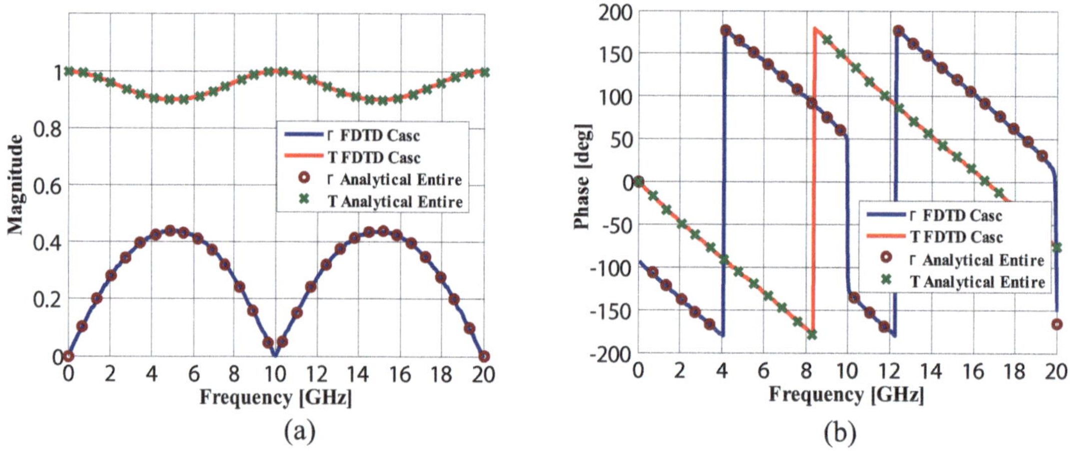

Figure 5.12: Reflection and transmission coefficients of infinite dielectric slab with normal incidence, (a) Magnitude, (b) Phase.

do not exist even for a very small gap (zero gap), and only the dominant mode is considered in the analysis.

From Figs. 5.12, 5.13, and 5.14, it should be noticed that the cascading technique is very accurate in calculating the S-parameters of the entire structure. In addition, good agreement can be noticed between the proposed technique and the analytical solution for both magnitude and phase of the reflection and transmission coefficients, with both oblique and normal incidence TE^z and TM^z cases.

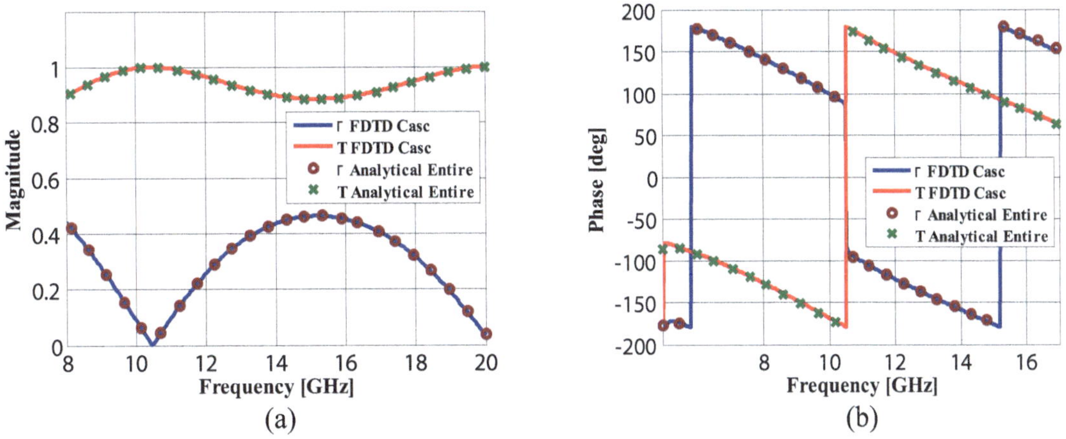

Figure 5.13: Reflection and transmission coefficients of infinite dielectric slab with oblique incidence $k_x = 104.8\ \mathrm{m}^{-1}$ TE^z, (a) Magnitude, (b) Phase.

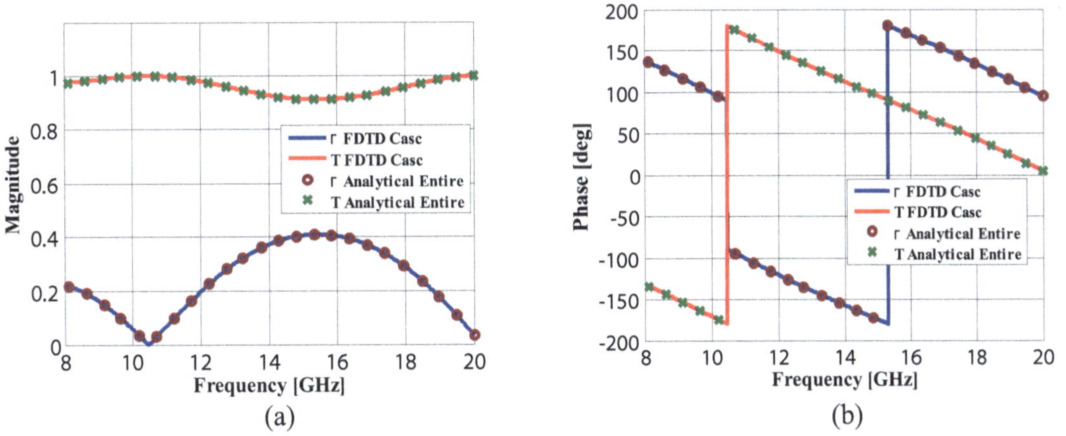

Figure 5.14: Reflection and transmission coefficients of infinite dielectric slab with oblique incidence $k_x = 104.8\ \mathrm{m}^{-1}$ TM^z, (a) Magnitude, (b) Phase.

5.5.2 TEST CASE 2 (1:1 CASE, NORMAL INCIDENCE AND LARGE GAP)

In this test case, the multi-layer geometry consists of two identical FSS structures consisting of dipole elements (1:1 case) separated by an air gap of width d. The dipole length is 12 mm and width is 3 mm. The periodicity is 15 mm in both x- and y-directions. The substrate has a thickness of 6 mm and relative permittivity $\varepsilon_r = 2.2$, as shown in Fig. 5.15. The structure is illuminated by a TE^z

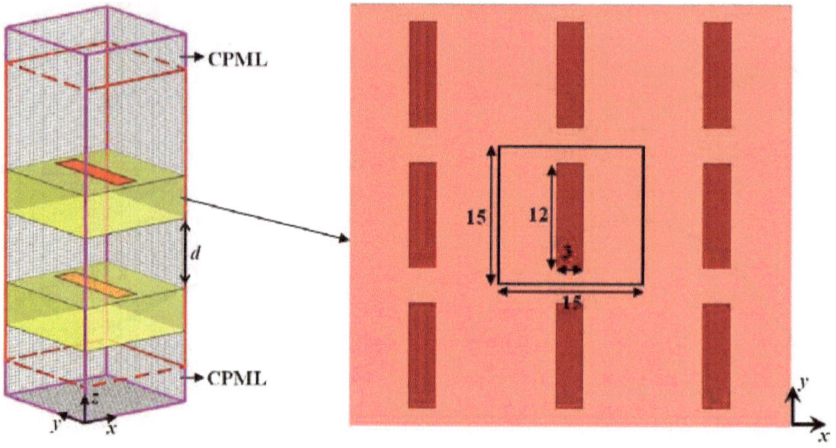

Figure 5.15: Two identical dipole FSS geometry (all dimensions are in mm).

normally incident plane wave (with polarization along y-axis). The frequency range of interest is 0-16 GHz. The FDTD grid cell size is $\Delta x = \Delta y = \Delta z = 0.5$ mm and 2,500 time steps and a 0.9 reduction factor of CFL time step are used. The CPML is used as the absorbing boundaries at the top and the bottom of the computational domain. The first step is to determine the distance d after which the level of all the harmonics are less than -40 dB relative to the corresponding magnitude of the incident field components. Using the gap determination procedure:

1) The two layers are identical; analyzing the harmonics of one layer is enough. The reflection and transmission harmonics must be calculated.

2) The frequency range of interest as specified by the problem is 0-16 GHz (as shown in Fig. 5.9, at the highest frequency the effect of harmonics is maximum).

3) k_x^i and k_y^i are equal to zero (normal incidence). Determine the cut-off frequencies for the first eight harmonics as follows:

$$
\begin{aligned}
M_{0,1}\,, M_{0,-1} &\rightarrow f_{cut\text{-}off}^{0,1} = f_{cut\text{-}off}^{0,-1} = 20\text{GHz} \\
M_{1,0}\,, M_{-1,0} &\rightarrow f_{cut\text{-}off}^{1,0} = f_{cut\text{-}off}^{-1,0} = 20\text{GHz} \\
M_{1,1}\,, M_{1,-1} &\rightarrow f_{cut\text{-}off}^{1,1} = f_{cut\text{-}off}^{1,-1} = 28.3\text{GHz} \\
M_{-1,1}\,, M_{-1,-1} &\rightarrow f_{cut\text{-}off}^{-1,1} = f_{cut\text{-}off}^{-1,-1} = 28.3\text{GHz}
\end{aligned}
$$

4) Use the harmonic analysis to calculate the magnitude coefficient of the eight harmonics and plot the behavior of these harmonics versus frequency, as shown in Figs. 5.16 and 5.17.

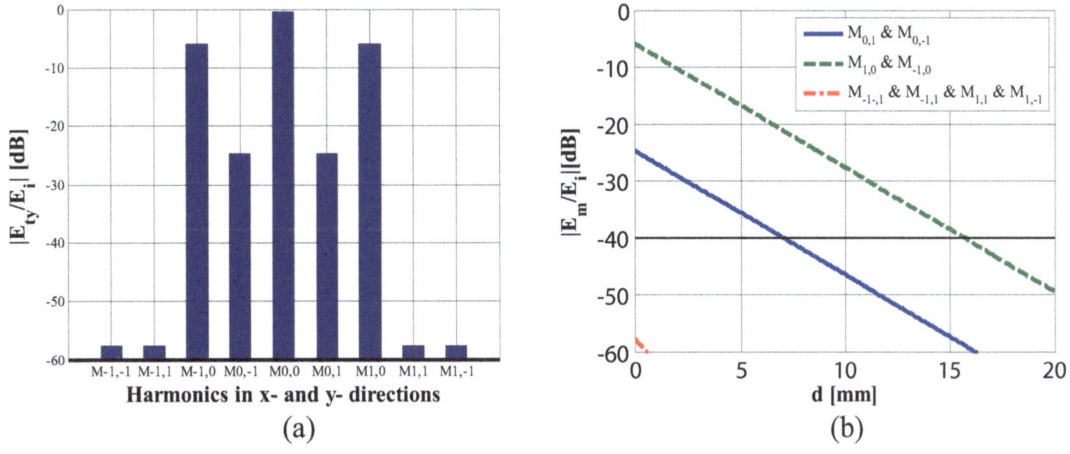

Figure 5.16: The eight transmitted harmonics at 16 GHz: (a) Magnitude compared to incident electric field, (b) Decaying relative magnitude versus gap distance.

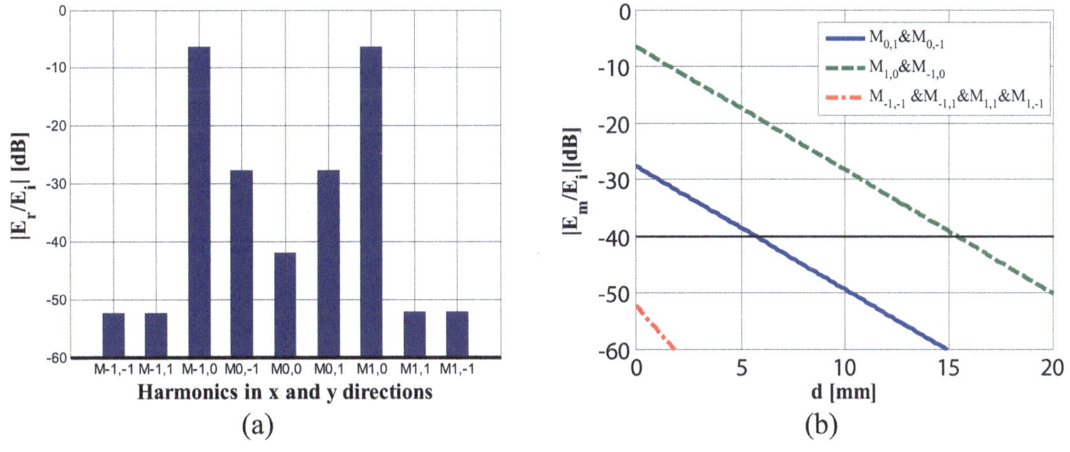

Figure 5.17: The eight reflected harmonics at 16 GHz: (a) Magnitude compared to incident electric field, (b) Decaying relative magnitude versus gap distance.

As can be noticed from Figs. 5.16 and 5.17, almost 95% of the dominant mode will be transmitted. In addition, a distance $d = 15.5$ mm between the two layers for this range of frequencies is considered enough to neglect all the higher harmonics effects (the magnitude of all higher harmonics are less than -40 dB compared to the incident field magnitude). To validate the cascading technique and the gap determination procedure, several air gap distances are analyzed and compared with the FDTD simulation of the entire structure, as shown in Fig. 5.18.

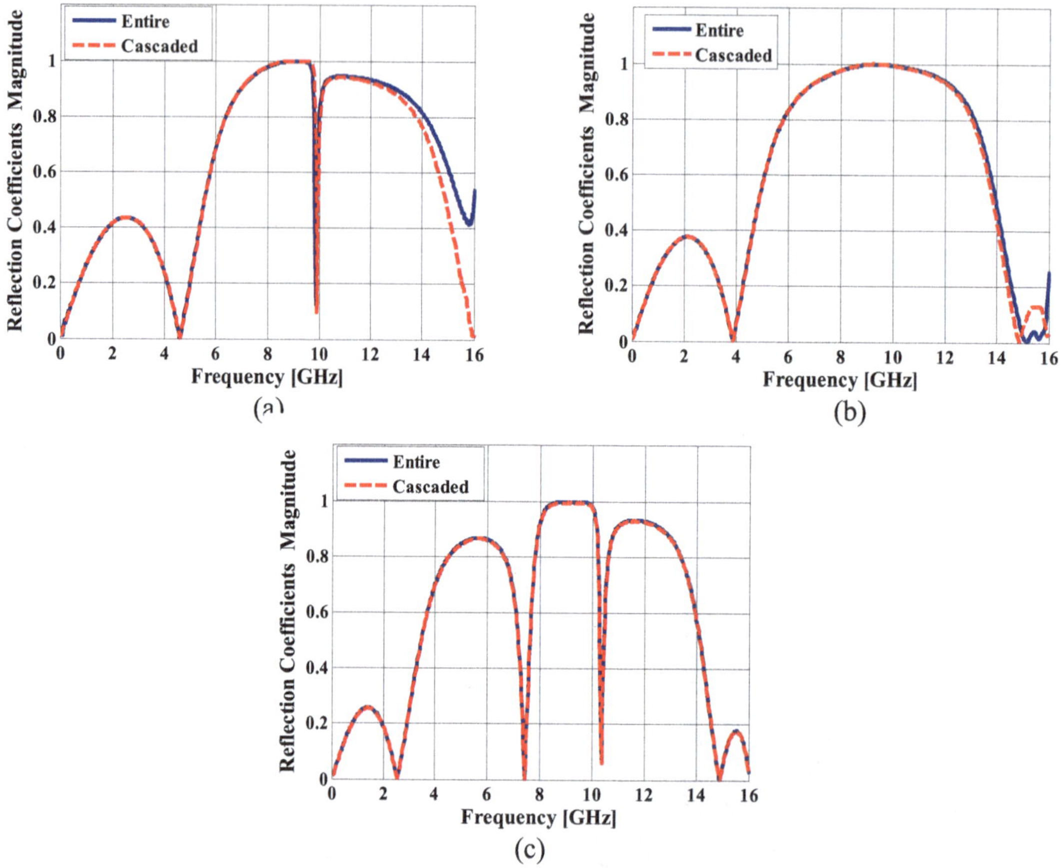

Figure 5.18: Reflection coefficients of two identical dipole FSS with normal incidence for (a) $d = 4$ mm, (b) $d = 7$ mm, (c) $d = 17$ mm (from [59]).

It should be noticed from Fig. 5.19, that when the gap size d is less than 15.5 mm, the cascading technique using only the dominant mode is not accurate, especially at high frequency, which validates the gap determination procedure. The relative error was calculated as follows:

$$error(f) = \frac{||\Gamma_{entire}(f)| - |\Gamma_{cascaded}(f)||}{\max(|\Gamma_{entire}(f)|)} \times 100\%. \tag{5.16}$$

The maximum relative error in case of a 4 mm gap is about 52%, and for $d = 7$ mm it is about 23% due to neglecting the higher harmonics effects which cause the frequency shift noticed in

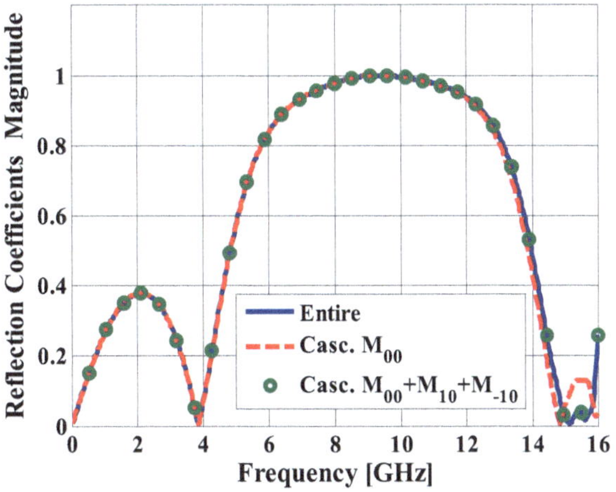

Figure 5.19: The reflection coefficients of two identical dipole FSS normal incident with $d = 7$ mm.

Fig. 5.19 (a) and (b). However, for the case of a gap size of 17 mm, the relative error is about 0.4%. The computational time using the cascading technique is less than the computational time for the entire structure, especially with large gaps (which require a large number of time steps to generate stable results). The computational time for the cascading case for d = 17 mm is 6 minutes (for calculating the S-parameters of the FSS layer and the total GSM), while the entire simulation for the same case takes 35 minutes, which demonstrates the efficiency of the hybrid FDTD/GSM technique. In addition, the domain size for the cascading case is equal to 43,200 cells (30 × 30 × 48), while for the entire structure, the domain size is 73,800 cells (30 × 30 × 82), which demonstrates the efficiency of the hybrid FDTD/GSM algorithm in terms of memory usage. Moreover, the scattering parameters generated for this layer can be saved and reused in any other cascading structure that uses the same layer with the same angle of incidence and frequency range (so the same S-parameters for the layer were used with the three gap sizes only the S-parameters of air gap where changed, while for the entire structure the whole simulation had to be repeated for each case).

To study the effect of the geometry on the harmonic frequency behavior, a test case is shown in Appendix B.2, where the two layers of the multi-layer periodic structure have the same periodicity as test case 2, but the elements are square patches instead of dipoles. In addition, another test case is shown in Appendix B.3, where the two layers of the multi-layer periodic structure have the same periodicity as test case 2, but the elements are L-shaped dipoles to study the effect of the cross-polarized field components.

5.5.3 TEST CASE 3 (1:1 CASE, NORMAL INCIDENCE AND SMALL GAP)

To analyze the same structure shown in Fig. 5.15 accurately with a gap size less than 15.5 mm, the cascading technique should include all the harmonics that have a magnitude greater than −40 dB compared to the incident (to achieve the required accuracy). For example, the case of gap size d = 7 mm is considered. From Figs. 5.16 and 5.17, it should be noticed that for a gap size of 7 mm, only two harmonics need to be added in the analysis to get accurate results ($M_{1,0}$ and $M_{-1,0}$). These two harmonics have a magnitude higher than −40 dB compared to the incident field at 7 mm. The S-parameters of these harmonics can be calculated from (5.8). Figure 5.19 compares the results of the cascading technique while using only the dominant mode and while using the dominant mode and the first two harmonics ($M_{1,0}$) and ($M_{-1,0}$).

It should be noticed that including the two harmonics in the cascading analysis improves the results. The small gap case can be easily analyzed after using harmonic analysis to determine exactly which harmonics should be considered in the analysis. The maximum relative error in the case of cascading technique with dominant mode and the two harmonics included is 0.5%.

5.5.4 TEST CASE 4 (1:1 CASE, OBLIQUE INCIDENCE AND LARGE GAP)

To study the effect of cross-polarization components, the algorithm is used to analyze the same structure shown in Fig. 5.15 with a general oblique incident plane wave k_x = 20 m^{-1} and k_y = 10 m^{-1}(general oblique incidence for minimum frequency of almost 1 GHz and angle φ = 26.65°). The frequency range of interest is 5-15 GHz. The FDTD grid cell size is $\Delta x = \Delta y = \Delta z = 0.5$ mm, 3,000 time steps and a Courant factor of 0.9 are used. Using the procedure of gap determination used in Section 5.4.2, different harmonics can be plotted versus distance at the highest frequency (15 GHz) in the frequency range of interest, as shown Fig. 5.20.

From Fig. 5.20, it can be concluded that d = 16.45 mm is large enough to neglect the effect of all the harmonics. In addition, it should be noticed that the case of oblique incidence required a larger gap to neglect the harmonics as compared to the case of normal incidence. The reflection and transmission coefficients of the whole structure are calculated using the cascading technique for different values of d, and the results are compared with the FDTD simulation of the entire structure, as shown in Fig. 5.21. It should be noticed from the figure that when d is less than 16.45 mm, inaccurate results are obtained from the cascading technique due to the effect of the harmonics; while when d is larger than 16.45 mm, accurate results are obtained. In addition, the oblique incidence will generate cross-polarized components, and these components must be considered in the analysis using Equation (5.4) as described in Section 5.3.2. The maximum relative error in the case of d =10 mm is about 8.2%, while it is about 1.2% in the case of d =18 mm. The computational time using the cascading technique is much less than the computational time for the entire structure, especially with large gaps.

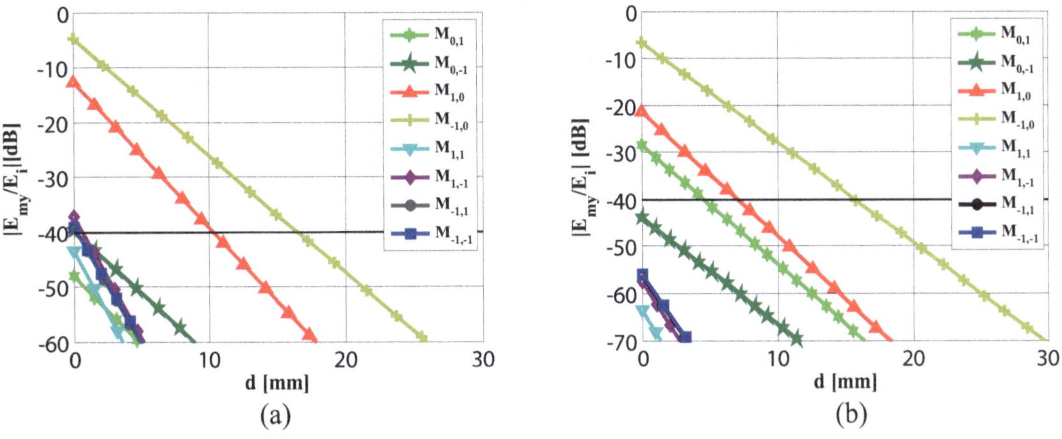

Figure 5.20: The first eight harmonics of dipole FSS layer at 15 GHz with oblique incidence ($k_x = 20\ \text{m}^{-1}$, $k_y = 10\ \text{m}^{-1}$), (a) Reflected components, (b) Transmitted components.

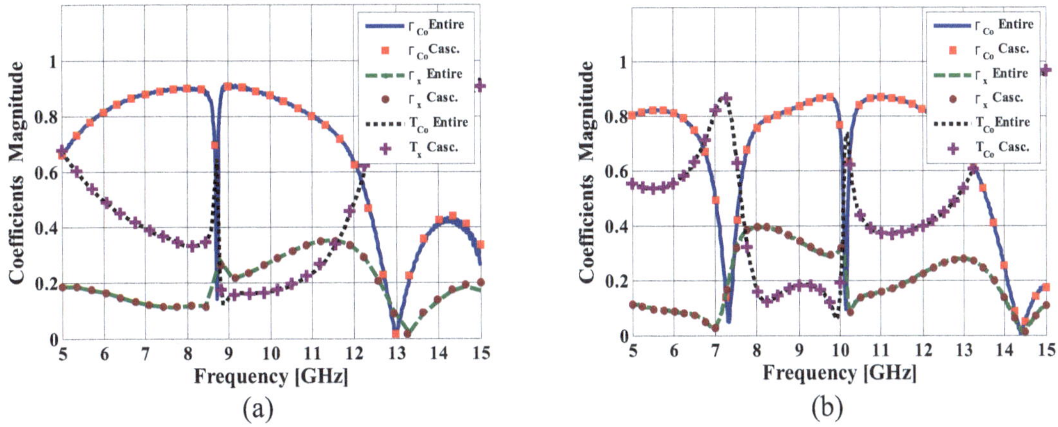

Figure 5.21: Reflection and transmission coefficients of two identical dipole FSS with oblique incidence TE^z case ($k_x = 20\ \text{m}^{-1}$, $k_y = 10\ \text{m}^{-1}$), (a) $d = 10$ mm, (b) $d = 18$ mm.

5.5.5 TEST CASE 5 (1:1 CASE, OBLIQUE INCIDENCE AND SMALL GAP)

The algorithm is used to analyze the same structure shown in Fig. 5.15. The structure is illuminated by an obliquely incident plane wave $k_x = 40\ \text{m}^{-1}$ and $k_y = 0\ \text{m}^{-1}$(for minimum frequency of almost 1.9 GHz); the frequency range of interest is 5-15 GHz, and $d = 10$ mm. The structure is to be simulated using the cascading technique; the same procedure used in test case 4 was used, and it was found that for a gap of 10 mm at a frequency of 15 GHz, only one harmonic needs to be added in the analysis to get accurate results from the cascading technique ($M_{-1,0}$). Figure 5.22 shows the results of the co-polarized reflection coefficient using the cascading technique with the dominant mode only and with the dominant mode plus the harmonic ($M_{-1,0}$). The results are compared with the FDTD simulation of the entire structure.

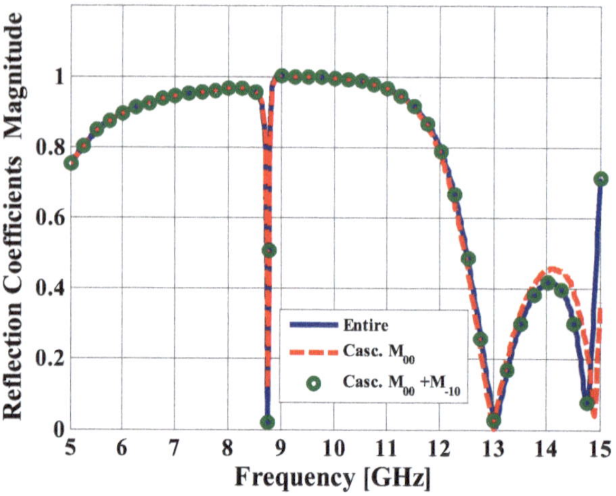

Figure 5.22: The reflection coefficient of two identical dipole FSS oblique incident ($k_x = 40\,\text{m}^{-1}$, $k_y = 0\,\text{m}^{-1}$), TE^z case with $d = 10$ mm.

It should be noticed from the Fig. 5.22 that when the effect of the ($M_{-1,0}$) harmonic is taken into consideration, accurate results are obtained. The maximum relative error in the case of cascading technique with only the dominant mode was calculated using (5.16) to be 38% (due to the frequency shift), while for the case in which the ($M_{-1,0}$) harmonic is included a maximum relative error of 0.6% is obtained.

5.5.6 TEST CASE 6 (N:M CASE, NORMAL INCIDENCE AND LARGE GAP)

As shown in Fig. 5.23, in this test case, the multi-layer geometry consists of two different FSS layers. The first FSS structure consists of square patch elements with a size of 6 mm. The periodicity is 10 mm in both the x- and y-directions. The substrate has a thickness of 6 mm and relative permittivity $\varepsilon_r = 2.2$. The second FSS structure is the same as the FSS structure used in test case 2 (Fig. 5.15) (general case n:m). The structure is illuminated by a normally incident plane wave ($k_x = k_y = 0\,\text{m}^{-1}$) and the frequency range of interest is 0-16 GHz. The FDTD grid cell size is $\Delta x = \Delta y = \Delta z = 0.5$ mm. 2,500 time steps and a 0.9 reduction factor of CFL time step are used. The first step is to determine the distance d after which all level of the harmonics becomes lower than -40 dB from the magnitude of the corresponding incident field component. Using the gap determination procedure described in Section 5.4.2 this distance can be easily determined. For the

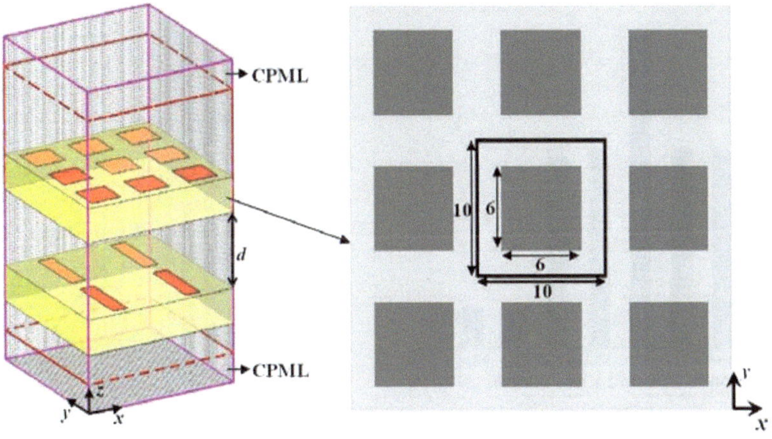

Figure 5.23: Square FSS and dipole FSS geometry n:m case (all dimensions are in mm) (from [60] © IEEE).

first layer only, the transmitted harmonics will affect the cascaded structure. As for the second layer, only the reflected harmonics will affect the cascaded structure.

1) The frequency range of interest as specified by the problem is 0-16 GHz.

2) k_x^i and k_y^i are equal to zero (normal incidence).

3) Determine the cut-off frequencies for the first eight harmonics of the first layer as follows:

$$M_{0,1}\,,\,M_{0,-1},\,M_{1,0}\,,\,M_{-1,0}(f_{cut\text{-}off}=30\text{GHz})$$
$$M_{1,1}\,,\,M_{1,-1},\,M_{-1,1}\,,\,M_{-1,-1}(f_{cut\text{-}off}=42.42\text{GHz})$$

The harmonics of the second layer are the same as in test case 2.

We use the harmonic analysis to calculate the magnitude coefficients of the first eight harmonics of layers 1 and 2 and plot the behavior of these harmonics with frequency as shown in Figs. 5.24 and 5.17. Figure 5.24 describes the transmitted harmonics from layer 1 at 16 GHz, while Fig. 5.17 describes the reflected harmonics from the second layer. It should be noticed from Fig. 5.24 that only 57% of the dominant mode will be transmitted from the first layer. In addition, due to the smaller periodicity compared to the second layer, the harmonics generated at the first layer will decay faster than the harmonics generated at the second layer. Using this information, it can be concluded that the second layer harmonics control the gap size.

To calculate the proper gap size after which the harmonics for both layers decay below −40 dB, the reflected harmonics shown in Fig. 5.17 should be multiplied by 0.57 because of the smaller incident wave illuminating the second layer, as shown in Fig. 5.24. A gap of 13.11 mm was found to be large enough to neglect the higher order harmonics effects.

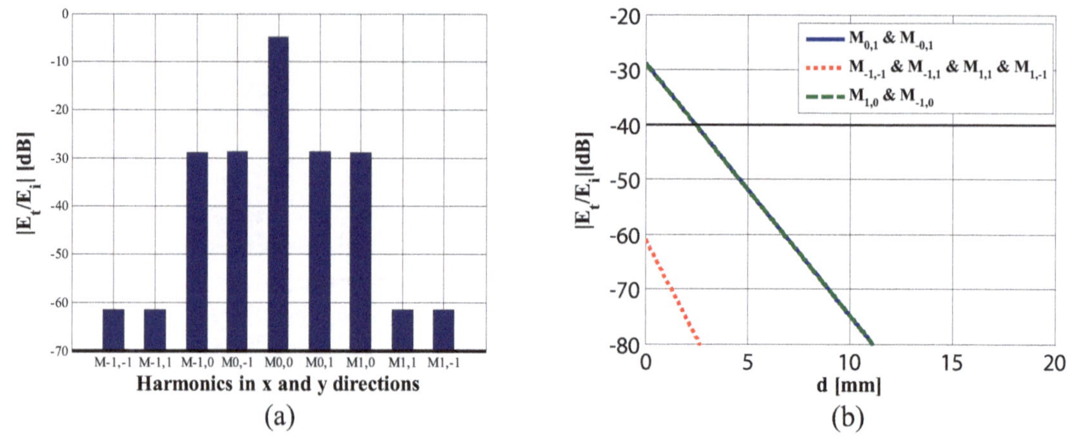

Figure 5.24: The first four transmitted harmonics from layer 1 at 16 GHz, (a) Magnitude compared to incident electric field, (b) Decaying relative magnitude versus gap distance.

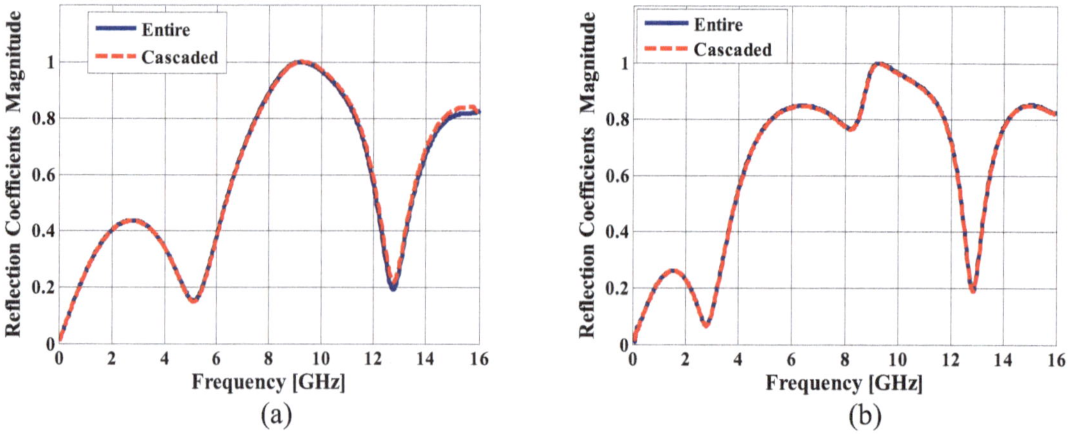

Figure 5.25: The reflection coefficient of square patch FSS and dipole FSS with normal incidence TE^z case, (a) $d = 3.5$ mm, and (b) 15 mm. (from [60] © IEEE).

To validate the cascading technique and the gap determination procedure, two air gaps were analyzed using the cascading technique (in the cascading technique only one unit cell from each layer is analyzed) and then comparing it with the FDTD simulation of the entire structure, as shown in Fig. 5.25. The maximum relative error in the case of d = 3.5 mm is 3%, while in the case of d = 15 mm, it is less than 0.3%. This test case is less sensitive for the harmonic effect, which might be due to the high cut-off frequencies of the harmonics generated from the first layer compared to the second layer. The computational time using the cascading technique is less than the computational time for the entire structure, especially with large gaps. In addition, to simulate the entire structure, many unit cells are needed for each layer. However, by using the cascading technique, only one unit cell is simulated for each layer, which reduces the computational time dramatically. The computational time for the cascading case is 8 minutes (for calculating S-parameters of the FSS layers and the total GSM), while for the simulation of the entire structure, it takes 130 minutes. Moreover, the domain size for the cascading case is equal to 43,200 cells ($30 \times 30 \times 48$), while for the entire structure, the domain size is 280,800 cells ($60 \times 60 \times 78$), which demonstrates the efficiency of the hybrid FDTD/GSM algorithm in terms of the memory usage. In addition, the entire structure simulation requires a large number of time steps to generate stable results.

5.5.7 TEST CASE 7 (N:M CASE, NORMAL INCIDENCE AND SMALL GAP)

To study the same structure shown in Fig. 5.23 with a small gap, the algorithm is used to analyze the structure with a gap size equal to 3.5 mm. From Fig. 5.24, it should be noticed that all the higher harmonics transmitted from the first layer will reach −40 dB at a distance of 2.5 mm, so for the gap size of 3.5 mm, the harmonics of the first layer can be neglected. For the second layer, all the harmonics of Fig. 5.17 (b) should be multiplied by 0.57. It was found that for a gap size of 3.5 mm, only two harmonics of the second layer are required to be added in the analysis to get accurate results from the cascading technique ($M_{1,0}$, $M_{-1,0}$ of the second layer). As long as three modes are included in the analysis, the S-matrix of each layer will be of the size 6×6 elements. To calculate the S-parameters of each layer, Equation (5.8) is used with the following parameters: E^{Dom} is related to the dominant mode ($k_x = k_y = 0\ \mathrm{m}^{-1}$), E^{Harm1} is related to the first harmonic of the second layer ($M_{1,0}$, $k_x = 418.879\ \mathrm{m}^{-1}$ and $k_y = 0\ \mathrm{m}^{-1}$), and E^{Harm2} is related to the second harmonic of the second layer ($M_{-1,0}$, $k_x = -418.879\ \mathrm{m}^{-1}$ and $k_y = 0\ \mathrm{m}^{-1}$). Thus, calculate S_{12} of the first layer, for example, the layer should be excited with the first harmonic of the second layer ($M_{1,0}$). Similarly, all other S-parameters of the two layers can be calculated.

Figure 5.26 shows the results of the co-polarized reflection coefficient using the cascading technique with only the dominant mode and with the dominant mode plus the harmonics ($M_{1,0}$) and ($M_{-1,0}$) of the second layer. The results are compared with the FDTD simulation of the entire structure. It should be noticed from Fig. 5.26 that when the effect of the ($M_{1,0}$) and ($M_{-1,0}$) harmonics are taken into consideration, accurate results are obtained. The maximum relative error in the case of the cascading technique with only the dominant mode was calculated using (5.16) to be 3%, while the case of the harmonics ($M_{1,0}$) and ($M_{-1,0}$) included results in a maximum relative error of 0.3%.

5.5.8 TEST CASE 8 (N:M CASE, OBLIQUE INCIDENCE AND LARGE GAP)

The algorithm is then used to analyze the same structure shown in Fig. 5.23, which is illuminated by an oblique incidence plane wave $k_x = 20\ \mathrm{m}^{-1}$ and $k_y = 10\ \mathrm{m}^{-1}$ (general oblique incident for minimum frequency of almost 1 GHz and angle $\varphi = 26.65°$). Using the procedure of gap determination, one

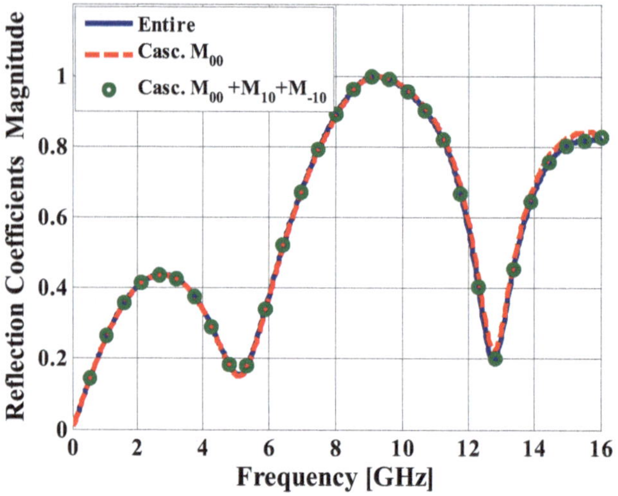

Figure 5.26: The reflection coefficient of square patch FSS and dipole FSS with normal incidence TE^z case with $d = 3.5$ mm (from [60] © IEEE).

can determine the gap distance d after which all levels of the harmonics reach -40 dB relative to the magnitude of the corresponding incident field components. For the first layer, only the transmitted harmonics will affect the cascaded structure; as for the second layer, only the reflected harmonics will affect the cascaded structure. Using the gap determination procedure:

1) The frequency range of interest as specified by the problem is 5-15 GHz.

2) $k_x^i = 20\ \mathrm{m}^{-1}$ and $k_y^i = 10\ \mathrm{m}^{-1}$ (General oblique incidence).

3) Determine the cut-off frequencies for the first four harmonics for the first layer and second layer as follows:

$$S_1 : f_{cut\text{-}off}^{1,0}{=}30.9\text{GHz},\ f_{cut\text{-}off}^{-1,0}{=}29\text{GHz},\ f_{cut\text{-}off}^{0,1}{=}30.5\text{GHz},\ f_{cut\text{-}off}^{0,-1}{=}29.5\text{GHz},$$

$$S_2 : f_{cut\text{-}off}^{1,0}{=}20.9\text{GHz},\ f_{cut\text{-}off}^{-1,0}{=}19.1\text{GHz},\ f_{cut\text{-}off}^{0,1}{=}20.5\text{GHz},\ f_{cut\text{-}off}^{0,-1}{=}19.5\text{GHz}.$$

4) Use the harmonic analysis to calculate the magnitudes of the first eight harmonics of layers 1 and 2. The behavior of these harmonics with frequency is plotted in Figs. 5.27 and 5.28 (E_x and E_y component).

5) Set -40 dB as the threshold for neglecting the effect of the harmonic effect.

Figure 5.27 describes the transmitted harmonics from the first layer at 15 GHz, while Fig. 5.28 describes the reflected harmonics from the second layer. A gap distance d = 14.95 mm was found to be large enough to neglect all the higher harmonics effects.

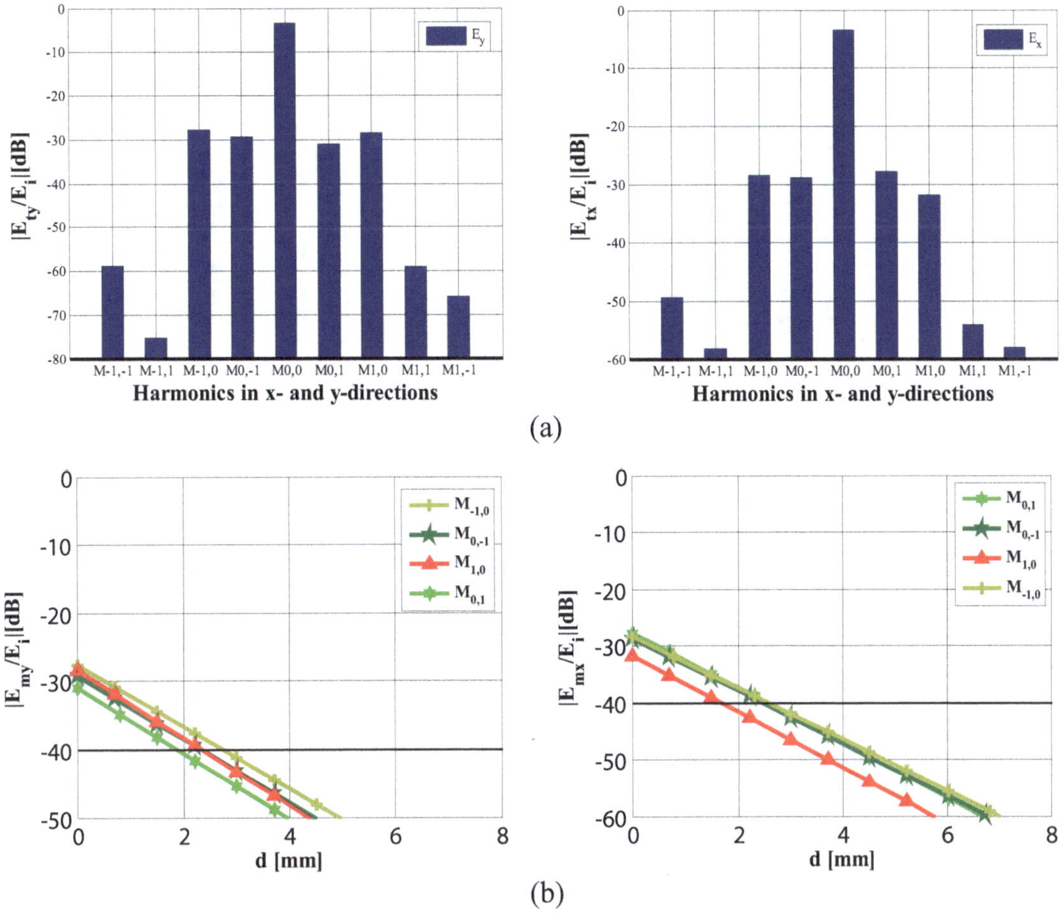

Figure 5.27: The transmitted harmonics from first layer at 15 GHz, (a) Magnitude of first eight harmonics compared to incident electric field, (b) The decaying of first four harmonics with distance.

The structure is analyzed using the cascading technique (only one unit cell from each layer is analyzed) with d = 15 mm and compared with the FDTD simulation of the entire structure, as shown in Fig. 5.29. The maximum error in the case of d = 15 mm is about 0.47%. The computational time using the cascading technique is less than the computational time for the entire structure.

5.6 SUMMARY

In this chapter, an efficient hybrid FDTD/GSM technique is described. In this technique, the constant horizontal wavenumber FDTD/PBC approach is used to compute the scattering parameters of

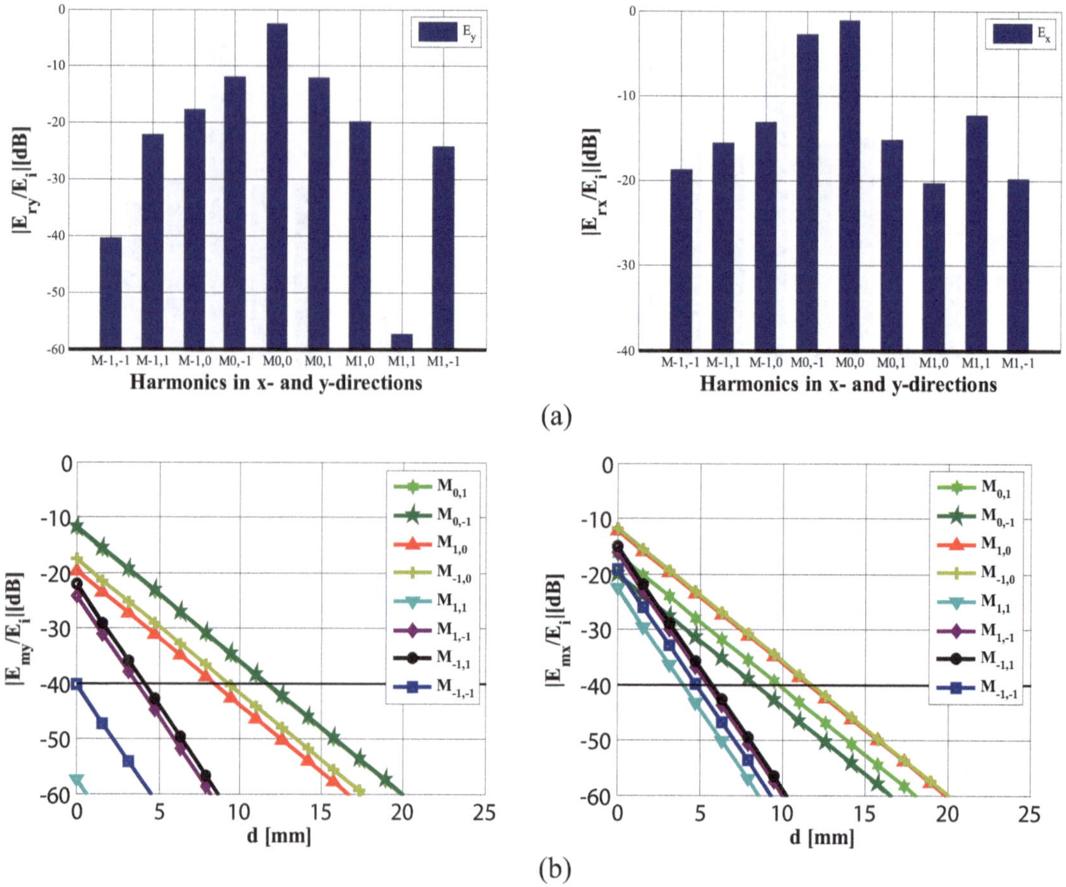

Figure 5.28: The first eight reflected harmonics from second layer at 15 GHz, (a) Magnitude compared to incident electric field (b) The decaying with distance.

each layer, after which the scattering matrix of the entire structure is calculated using the cascading technique. In addition, two procedures were described: one is used to study the behavior of different harmonics (evanescent and propagating) using the constant horizontal wavenumber FDTD/PBC approach, which dramatically reduces memory usage; the other procedure is used to determine the proper gap size (for neglecting the harmonics effects), and it can also be used to select proper harmonics for a specific gap size. The validity of the algorithm was verified through several numerical examples including FSS structures with different periodicities and under different incident angles. The numerical results of the developed approach show good agreement with the results obtained from the direct FDTD simulation of the entire structure, while the proposed approach significantly saves computational time and memory usage.

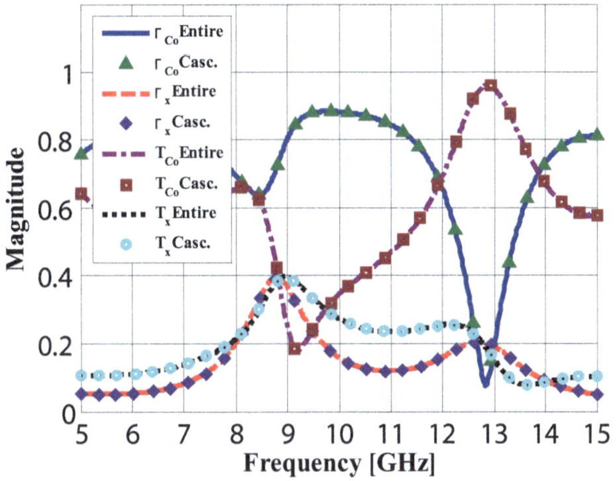

Figure 5.29: Reflection and transmission coefficients of square patch FSS and dipole FSS with oblique incidence (k_x = 20 m^{-1}, k_y = 10 m^{-1}) TE^z case for d = 15 mm.

CHAPTER 6

Conclusions

Periodic structures are of great importance in electromagnetics due to their wide range of applications. In this book various configurations with different material types of electromagnetic periodic structures such as multi-layered structures and arbitrary skewed grids with and without dispersive materials were analyzed using the constant horizontal wavenumber FDTD/PBC technique. A full description of the FDTD/PBC constant horizontal wavenumber approach was provided in Chapter 2. The FDTD updating equations were derived, and numerical results were provided to demonstrate the validity, advantages, and limitations of this approach.

In Chapter 3, the FDTD/PBC approach was expanded to analyze the scattering properties of general skewed-grid periodic structures. It is capable of calculating the co- and cross-polarization reflection coefficients for normal and oblique incidence with either TE^z or TM^z wave polarizations. The approach is simple and its stability criterion is independent from the incident wave angle, thus configurations with wave incident angles close to grazing can be considered. The numerical results showed good agreement with results from the analytical solution for a dielectric slab, and results from the MoM solutions for dipole and JC FSSs.

In Chapter 4, a new FDTD/DPBC approach to analyze the scattering properties of general periodic structures with dispersive material was introduced. The approach is developed based on both the constant horizontal wavenumber and the auxiliary differential equation (ADE) techniques. It is capable of calculating the co- and cross-polarization reflection and transmission coefficients, for both normal and oblique wave incidence, and for TE^z and TM^z wave polarizations. The numerical results show good agreement with results from the analytical solution for a water dispersive slab, and results from the frequency-domain solutions for periodic dispersive structures.

In Chapter 5, an efficient hybrid FDTD/GSM technique is described. In this technique the constant horizontal wavenumber FDTD/PBC approach is used to compute the scattering parameters of each layer, after which the scattering matrix of the entire structure is calculated using the cascading technique. In addition, two procedures were described; one is used to study the behavior of different harmonics (evanescent and propagating) using the constant horizontal wavenumber FDTD/PBC approach, which dramatically reduces the memory usage. The other procedure is used to determine the proper gap size (for neglecting the harmonics effects) and it can also be used to select the proper harmonics for a specific gap size. The validity of the algorithm was verified through several numerical examples including FSSs with different periodicities while being illuminated at different incident angles. The numerical results of the developed approach show good agreement with the results obtained from the direct FDTD simulation of the entire structure. This approach provides a comprehensive study for determining the proper gap size at which the higher order harmonics effects can be neglected. It also provides useful means to select the proper harmonics for the small gap of multilayered configurations. One of the main advantages of this approach is the capability to analyze multilayered periodic structures where each layer has its own unique element type, periodicity, and material decomposition can be easily conducted by this approach.

The algorithms developed in this book are implemented using MATLAB and for all the examples presented here, the developed MATLAB codes are found to be faster and more efficient in memory usage than traditional commercial software packages. These codes lead to comprehensive software tools that are capable of efficiently and accurately analyzing electromagnetic periodic structures of different configurations and material types and with the possibility of including linear and non-linear lumped circuit elements.

APPENDIX A

Dispersive Media

A.1 AUXILIARY DIFFERENTIAL EQUATION IN SCATTERED FIELD FORMULATION

For an FDTD scattered field formulation for dispersive media, the equations are as follows for plane wave excitation [42]:

$$\begin{aligned}
&\vec{D}(t) + (\tau_1 + \tau_2)\frac{\partial \vec{D}(t)}{\partial t} + \tau_1\tau_2\frac{\partial^2 \vec{D}(t)}{\partial t^2} \\
&= \varepsilon_o\left[\varepsilon_s \vec{E}_s(t) + (\varepsilon_{s1}\tau_2 + \varepsilon_{s2}\tau_1)\frac{\partial \vec{E}_s(t)}{\partial t} + \tau_1\tau_2\varepsilon_\infty\frac{\partial^2 \vec{E}_s(t)}{\partial t^2}\right] \\
&+ \varepsilon_o\left[\varepsilon_s \vec{E}_i(t) + (\varepsilon_{s1}\tau_2 + \varepsilon_{s2}\tau_1)\frac{\partial \vec{E}_i(t)}{\partial t} + \tau_1\tau_2\varepsilon_\infty\frac{\partial^2 \vec{E}_i(t)}{\partial t^2}\right],
\end{aligned} \tag{A.1}$$

$$\frac{\partial \vec{H}_s(t)}{\partial t} = -\frac{1}{\mu}\nabla \times \vec{E}_s, \tag{A.2}$$

$$\frac{\partial \vec{D}(t)}{\partial t} = \nabla \times \vec{H}_s + \varepsilon_0\frac{\partial \vec{E}_i(t)}{\partial t}, \text{ where } \nabla \times \vec{H}_i \to \varepsilon_0\frac{\partial \vec{E}_i(t)}{\partial t}. \tag{A.3}$$

Now it is obvious that the vector differential Equations (A.1)–(A.3) are expressed in terms of incident and scattered fields. The incident field and its derivatives are usually defined analytically. Using the central difference approximation for the derivatives in Equations (A.1)– (A.3), the updating equations for the components of the field vectors $\boldsymbol{E}_s, \boldsymbol{H}_s$, and the auxiliary displacement vector $\boldsymbol{D}$ can be easily obtained.

Similar to the total field formulation, the updating equations can be written in the same manner as in [1], assuming σ^m and $M_i = 0$:

$$\frac{\partial H_x^s}{\partial t} = \frac{1}{\mu_x}\left[\frac{\partial E_y^s}{\partial z} - \frac{\partial E_z^s}{\partial y}\right], \tag{A.4a}$$

$$\frac{\partial H_y^s}{\partial t} = \frac{1}{\mu_y}\left[\frac{\partial E_z^s}{\partial x} - \frac{\partial E_x^s}{\partial z}\right], \tag{A.4b}$$

$$\frac{\partial H_z^s}{\partial t} = \frac{1}{\mu_z}\left[\frac{\partial E_x^s}{\partial y} - \frac{\partial E_y^s}{\partial x}\right], \tag{A.4c}$$

$$\frac{\partial D_x}{\partial t} = \left[\frac{\partial H_z^s}{\partial y} - \frac{\partial H_y^s}{\partial z}\right] + \varepsilon_0 \frac{\partial E_x^i}{\partial t}, \tag{A.4d}$$

$$\frac{\partial D_y}{\partial t} = \left[\frac{\partial H_x^s}{\partial z} - \frac{\partial H_z^s}{\partial x}\right] + \varepsilon_0 \frac{\partial E_y^i}{\partial t}, \tag{A.4e}$$

$$\frac{\partial D_z}{\partial t} = \left[\frac{\partial H_y^s}{\partial x} - \frac{\partial H_x^s}{\partial y}\right] + \varepsilon_0 \frac{\partial E_z^i}{\partial t}, \tag{A.4f}$$

$$\begin{aligned} D_x + [\tau_1 + \tau_2]\frac{\partial D_x}{\partial t} + \tau_1\tau_2\frac{\partial^2 D_x}{\partial t^2} &= \varepsilon_o\varepsilon_s[E_x^s + E_x^i] + \varepsilon_o[\varepsilon_{s1}\tau_2 + \varepsilon_{s2}\tau_1]\left[\frac{\partial E_x^s}{\partial t} + \frac{\partial E_x^i}{\partial t}\right] \\ &+ \tau_1\tau_2\varepsilon_o\varepsilon_\infty\left[\frac{\partial^2 E_x^s}{\partial t^2} + \frac{\partial^2 E_x^i}{\partial t^2}\right], \end{aligned} \tag{A.4g}$$

$$\begin{aligned} D_y + [\tau_1 + \tau_2]\frac{\partial D_y}{\partial t} + \tau_1\tau_2\frac{\partial^2 D_y}{\partial t^2} &= \varepsilon_o\varepsilon_s[E_y^s + E_y^i] + \varepsilon_o[\varepsilon_{s1}\tau_2 + \varepsilon_{s2}\tau_1]\left[\frac{\partial E_y^s}{\partial t} + \frac{\partial E_y^i}{\partial t}\right] \\ &+ \tau_1\tau_2\varepsilon_o\varepsilon_\infty\left[\frac{\partial^2 E_y^s}{\partial t^2} + \frac{\partial^2 E_y^i}{\partial t^2}\right], \end{aligned} \tag{A.4h}$$

$$\begin{aligned} D_z + [\tau_1 + \tau_2]\frac{\partial D_z}{\partial t} + \tau_1\tau_2\frac{\partial^2 D_z}{\partial t^2} &= \varepsilon_o\varepsilon_s[E_z^s + E_z^i] + \varepsilon_o[\varepsilon_{s1}\tau_2 + \varepsilon_{s2}\tau_1]\left[\frac{\partial E_z^s}{\partial t} + \frac{\partial E_z^i}{\partial t}\right] \\ &+ \tau_1\tau_2\varepsilon_o\varepsilon_\infty\left[\frac{\partial^2 E_z^s}{\partial t^2} + \frac{\partial^2 E_z^i}{\partial t^2}\right]. \end{aligned} \tag{A.4i}$$

By re-arranging the above nine equations the recursive FDTD algorithm can be easily written, starting with H_x, E_x, and D_x as follows [1]:

For the H_x component:

$$\begin{aligned} H_x^{n+\frac{1}{2}}(i,j,k) = &C_{hxh}(i,j,k) \times H_x^{n-\frac{1}{2}}(i,j,k) \\ &+ C_{hxey}(i,j,k) \times \left[E_y^n(i,j,k+1) - E_y^n(i,j,k)\right] \\ &+ C_{hxez}(i,j,k) \times \left[E_z^n(i,j+1,k) - E_z^n(i,j,k)\right], \end{aligned} \tag{A.5}$$

where

$$C_{hxh}(i,j,k) = 1, C_{hxey}(i,j,k) = \frac{\Delta t}{(\mu_x(i,j,k))\Delta z}, C_{hxez}(i,j,k) = \frac{-\Delta t}{(\mu_x(i,j,k))\Delta y}.$$

For the H_y component:

$$\begin{aligned} H_y^{n+\frac{1}{2}}(i,j,k) = &C_{hyh}(i,j,k) \times H_y^{n-\frac{1}{2}}(i,j,k) \\ &+ C_{hyez}(i,j,k) \times \left[E_z^n(i+1,j,k) - E_z^n(i,j,k)\right] \\ &+ C_{hyex}(i,j,k) \times \left[E_x^n(i,j,k+1) - E_x^n(i,j,k)\right], \end{aligned} \tag{A.6}$$

where

$$C_{hyh}(i,j,k) = 1, C_{hyez}(i,j,k) = \frac{\Delta t}{(\mu_y(i,j,k))\Delta x}, C_{hyex}(i,j,k) = \frac{-\Delta t}{(\mu_y(i,j,k))\Delta z}.$$

For the H_z component:

$$\begin{aligned} H_z^{n+\frac{1}{2}}(i,j,k) = {} & C_{hzh}(i,j,k) \times H_z^{n-\frac{1}{2}}(i,j,k) \\ & + C_{hzex}(i,j,k) \times \left[E_x^n(i,j+1,k) - E_x^n(i,j,k)\right] \\ & + C_{hxey}(i,j,k) \times \left[E_y^n(i+1,j,k) - E_y^n(i,j,k)\right], \end{aligned} \tag{A.7}$$

where

$$C_{hzh}(i,j,k) = 1, C_{hzex}(i,j,k) = \frac{\Delta t}{(\mu_z(i,j,k))\Delta y}, C_{hzey}(i,j,k) = \frac{-\Delta t}{(\mu_z(i,j,k))\Delta x}.$$

As long as the μ (permeability) of the material is independent of the frequency, the updating equations for the magnetic field will be similar to the conventional FDTD updating equation. For updating the displacement field vector $\boldsymbol{D}$, we start by updating the D_x as follows:

$$\begin{aligned} D_x^{n+1}(i,j,k) = {} & C_{dxd}(i,j,k) \times D_x^n(i,j,k) \\ & + C_{dxhz}(i,j,k) \times [H_z^{n+\frac{1}{2}}(i,j,k) - H_z^{n+\frac{1}{2}}(i,j-1,k)] \\ & + C_{dxhy}(i,j,k) \times [H_y^{n+\frac{1}{2}}(i,j,k) - H_y^{n+\frac{1}{2}}(i,j,k-1)] \\ & + C_{dxexi}(i,j,k) \times [E_{inc,x}^{n+1}(i,j,k) - E_{inc,x}^{n}(i,j,k)], \end{aligned} \tag{A.8}$$

where

$$C_{dxd}(i,j,k) = 1,\ C_{dxhz}(i,j,k) = \frac{\Delta t}{\Delta y}, C_{dxhy}(i,j,k) = \frac{-\Delta t}{\Delta z}, C_{dxexi}(i,j,k) = \varepsilon_0,$$

similarly, for the D_y:

$$\begin{aligned} D_y^{n+1}(i,j,k) = {} & C_{dyd}(i,j,k) \times D_y^n(i,j,k) \\ & + C_{dyhx}(i,j,k) \times [H_x^{n+\frac{1}{2}}(i,j,k) - H_x^{n+\frac{1}{2}}(i,j,k-1)] \\ & + C_{dyhz}(i,j,k) \times [H_z^{n+\frac{1}{2}}(i,j,k) - H_z^{n+\frac{1}{2}}(i-1,j,k)] \\ & + C_{dyeyi}(i,j,k) \times [E_{inc,y}^{n+1}(i,j,k) - E_{inc,y}^{n}(i,j,k)], \end{aligned} \tag{A.9}$$

where

$$C_{dyd}(i,j,k) = 1,\ C_{dyhx}(i,j,k) = \frac{\Delta t}{\Delta z}, C_{dyhz}(i,j,k) = \frac{-\Delta t}{\Delta x}, C_{dyeyi}(i,j,k) = \varepsilon_0,$$

and similarly, for the D_z:

$$\begin{aligned} D_z^{n+1}(i,j,k) = & C_{dzd}(i,j,k) \times D_z^n(i,j,k) \\ & + C_{dzhy}(i,j,k) \times [H_y^{n+\frac{1}{2}}(i,j,k) - H_y^{n+\frac{1}{2}}(i-1,j,k)] \\ & + C_{dzhx}(i,j,k) \times [H_x^{n+\frac{1}{2}}(i,j,k) - H_x^{n+\frac{1}{2}}(i,j-1,k)] \\ & + C_{dzezi}(i,j,k) \times [E_{inc,z}^{n+1}(i,j,k) - E_{inc,z}^{n}(i,j,k)], \end{aligned} \tag{A.10}$$

where

$$C_{dzd}(i,j,k) = 1,\ C_{dzhy}(i,j,k) = \frac{\Delta t}{\Delta x}, C_{dzhx}(i,j,k) = \frac{-\Delta t}{\Delta y}, C_{dzezi}(i,j,k) = \varepsilon_0$$

To update the electric field vector $\boldsymbol{E}$, we start by updating E_x as follows:

$$\begin{aligned} & \frac{1}{2}(D_x^{n+1} + D_x^n) + \frac{\tau_1+\tau_2}{\Delta t}(D_x^{n+1} - D_x^n) + \frac{\tau_1\tau_2}{(\Delta t)^2}(D_x^{n+1} - 2D_x^n + D_x^{n-1}) = \\ & \frac{\varepsilon_o\varepsilon_s}{2}(E_x^{n+1} + E_x^n) + \frac{\varepsilon_o(\varepsilon_{s1}\tau_2 + \varepsilon_{s2}\tau_1)}{\Delta t}(E_x^{n+1} - E_x^n) + \frac{\varepsilon_o\varepsilon_\infty\tau_1\tau_2}{(\Delta t)^2}(E_x^{n+1} - 2E_x^n + E_x^{n-1}) + \\ & \frac{\varepsilon_o\varepsilon_s}{2}(E_{inc,x}^{n+1} + E_{inc,x}^n) + \frac{\varepsilon_o(\varepsilon_{s1}\tau_2 + \varepsilon_{s2}\tau_1)}{\Delta t}(E_{inc,x}^{n+1} - E_{inc,x}^n) + \\ & \frac{\varepsilon_o\varepsilon_\infty\tau_1\tau_2}{(\Delta t)^2}(E_{inc,x}^{n+1} - 2E_{inc,x}^n + E_{inc,x}^{n-1}). \end{aligned}$$

Then,

$$\begin{aligned} [\alpha_0^x + \alpha_1^x + \alpha_2^x]E_x^{n+1} = & [-\alpha_0^x + \alpha_1^x + 2\alpha_2^x]E_x^n + [-\alpha_2^x]E_x^{n-1} \\ & - [\alpha_0^x + \alpha_1^x + \alpha_2^x]E_{inc,x}^{n+1} + [-\alpha_0^x + \alpha_1^x + 2\alpha_2^x]E_{inc,x}^n + [-\alpha_2^x]E_{inc,x}^{n-1} \\ & + [\beta_0^x + \beta_1^x + \beta_2^x]D_x^{n+1} + [\beta_0^x - \beta_1^x - 2\beta_2^x]D_x^n + [\beta_2^x]D_x^{n-1} \end{aligned}$$

$$\begin{aligned} E_x^{n+1}(i,j,k) = & - E_{inc,x}^{n+1}(i,j,k) + C_{exe1} \times [E_x^n(i,j,k) + E_{inc,x}^n(i,j,k)] \\ & + C_{exe2} \times [E_x^{n-1}(i,j,k) + E_{inc,x}^{n-1}(i,j,k)] + C_{exd1} \times D_x^{n+1}(i,j,k) \\ & + C_{exd2} \times D_x^n(i,j,k) + C_{exd3} \times D_x^{n-1}(i,j,k) \end{aligned} \tag{A.11}$$

$$C_{exe1} = \frac{[-\alpha_0^x + \alpha_1^x + 2\alpha_2^x]}{[\alpha_0^x + \alpha_1^x + \alpha_2^x]}, C_{exe2} = \frac{[-\alpha_2^x]}{[\alpha_0^x + \alpha_1^x + \alpha_2^x]}, C_{exd1} = \frac{[\beta_0^x + \beta_1^x + \beta_2^x]}{[\alpha_0^x + \alpha_1^x + \alpha_2^x]},$$

$$C_{exd2} = \frac{[\beta_0^x - \beta_1^x - 2\beta_2^x]}{[\alpha_0^x + \alpha_1^x + \alpha_2^x]}, C_{exd3} = \frac{[\beta_2^x]}{[\alpha_0^x + \alpha_1^x + \alpha_2^x]},$$

similarly, for the E_y:

$$\begin{aligned} E_y^{n+1}(i,j,k) = &- E_{inc,y}^{n+1}(i,j,k) + C_{eye1} \times [E_y^n(i,j,k) + E_{inc,y}^n(i,j,k)] \\ &+ C_{eye2} \times [E_y^{n-1}(i,j,k) + E_{inc,y}^{n-1}(i,j,k)] + C_{eyd1} \times D_y^{n+1}(i,j,k) \\ &+ C_{eyd2} \times D_y^n(i,j,k) + C_{eyd3} \times D_y^{n-1}(i,j,k) \end{aligned} \tag{A.12}$$

$$C_{eye1} = \frac{[-\alpha_0^y + \alpha_1^y + 2\alpha_2^y]}{[\alpha_0^y + \alpha_1^y + \alpha_2^y]}, C_{eye2} = \frac{[-\alpha_2^y]}{[\alpha_0^y + \alpha_1^y + \alpha_2^y]}, C_{eyd1} = \frac{[\beta_0^y + \beta_1^y + \beta_2^y]}{[\alpha_0^y + \alpha_1^y + \alpha_2^y]},$$

$$C_{eyd2} = \frac{[\beta_0^y - \beta_1^y - 2\beta_2^y]}{[\alpha_0^y + \alpha_1^y + \alpha_2^y]}, C_{eyd3} = \frac{[\beta_2^y]}{[\alpha_0^y + \alpha_1^y + \alpha_2^y]},$$

and similarly, for the E_z:

$$\begin{aligned} E_z^{n+1}(i,j,k) = &- E_{inc,z}^{n+1}(i,j,k) + C_{eze1} \times [E_z^n(i,j,k) + E_{inc,z}^n(i,j,k)] \\ &+ C_{eze2} \times [E_z^{n-1}(i,j,k) + E_{inc,z}^{n-1}(i,j,k)] + C_{ezd1} \times D_z^{n+1}(i,j,k) \\ &+ C_{ezd2} \times D_z^n(i,j,k) + C_{ezd3} \times D_z^{n-1}(i,j,k) \end{aligned} \tag{A.13}$$

$$C_{eze1} = \frac{[-\alpha_0^z + \alpha_1^z + 2\alpha_2^z]}{[\alpha_0^z + \alpha_1^z + \alpha_2^z]}, C_{eze2} = \frac{[-\alpha_2^z]}{[\alpha_0^z + \alpha_1^z + \alpha_2^z]}, C_{ezd1} = \frac{[\beta_0^z + \beta_1^z + \beta_2^z]}{[\alpha_0^z + \alpha_1^z + \alpha_2^z]},$$

$$C_{ezd2} = \frac{[\beta_0^z - \beta_1^z - 2\beta_2^z]}{[\alpha_0^z + \alpha_1^z + \alpha_2^z]}, C_{ezd3} = \frac{[\beta_2^y]}{[\alpha_0^z + \alpha_1^z + \alpha_2^z]}.$$

A.2 SCATTERING FROM 3-D DISPERSIVE OBJECTS

To check the validity of the ADE scattered field formulation, the formulation was developed using MATLAB code and a test case was executed. In this test case, the bistatic RCS of a water dispersive sphere was calculated, where the sphere has a radius of 10 cm. The parameters for water permittivity are obtained from [38] as ε_{s1} = 81, ε_{s2} = 1.8, ε_∞ =1.8, τ_1 = 9.4 × 10^{-12} and τ_2= 0. The FDTD grid cell size is $\Delta x = \Delta y = \Delta z = 0.75$ cm. In the FDTD code 20,000 time steps and a 0.9 reduction factor of CFL time step are used. The CPML is used for the absorbing boundaries of the computational domain, as shown in Fig. A.1, and the cube is excited using a Gaussian pulse. The results are compared with results obtained from HFSS, as shown in Fig. A.2. Good agreement is noticed between results generated by the FDTD method and the results generated using the HFSS package, which proves the validity of the scattered field formulation.

A.3 ANALYSIS OF RFID TAGS MOUNTED OVER HUMAN BODY TISSUE

Radio frequency identification (RFID) is becoming one of the popular systems in today's societies. A practical application for the RFID tags is to use them to track animals or sometimes people (children or seniors), but mounting these tags on human tissues will affect their performance due

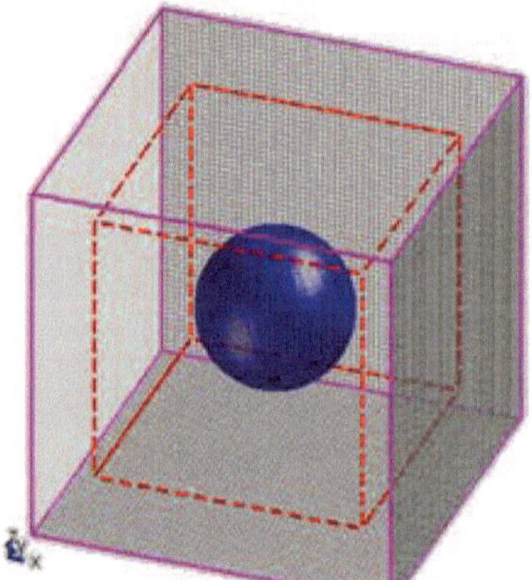

Figure A.1: Water dispersive sphere computational domain.

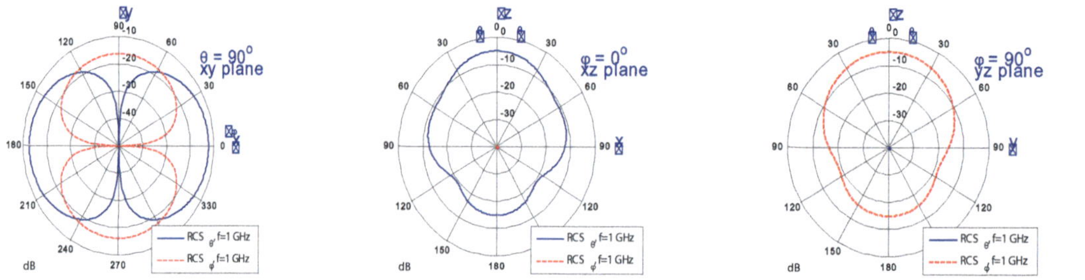

(a)

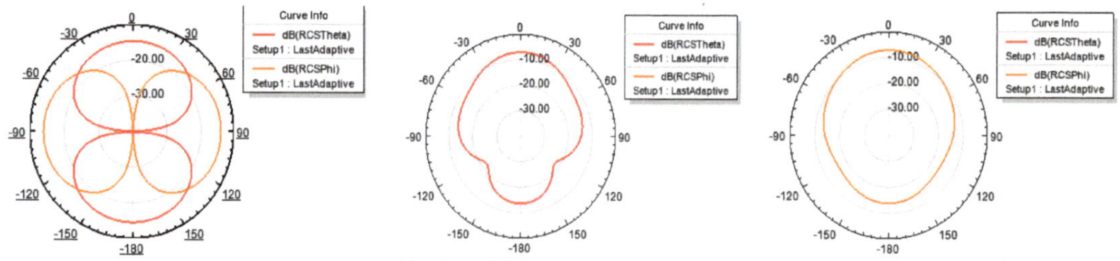

(b)

Figure A.2: Water dispersive sphere bistatic RCS at 1 GHz, (a) FDTD, (b) HFSS.

to the dispersive nature of the human tissues. In this section, three test cases are conducted to study the effect of human tissues on the performance of RFID tags. The geometry of the tag is shown in Fig. A.3, and the tag is designed to operate at 2.45 GHz. The details of the three test cases are shown in Table A.1.

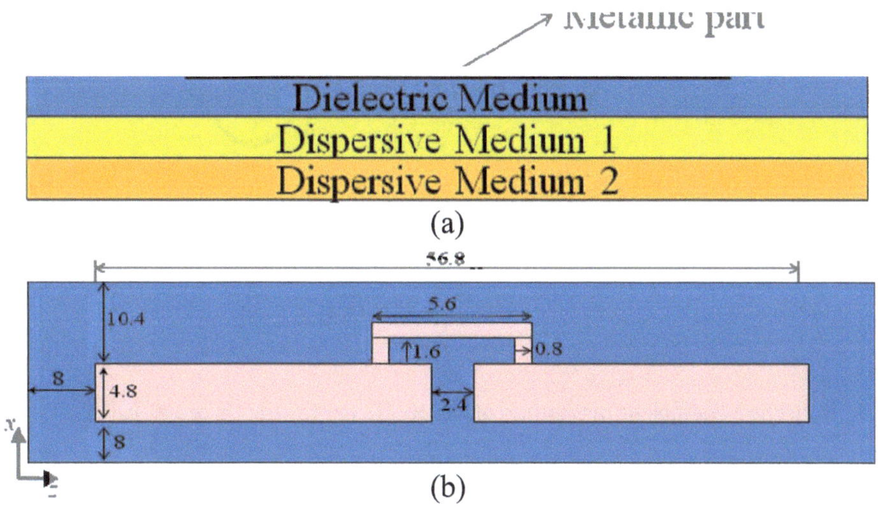

Figure A.3: Geometry of RFID tag mounted over a three medium substrate, (a) Side view, (b) Top view (all dimensions are in mm).

Table A.1: RFID tag test plan.

Test Case Number	Material 1	Material 2	Material 3
1	d_1 = 5mm, dielectric $\varepsilon_r = 2$	N/A	N/A
2	d_1 = 5mm, dielectric $\varepsilon_r = 2$	d_2 = 5mm, Dry Skin	N/A
3	d_1 = 5mm, dielectric $\varepsilon_r = 2$	d_2 = 5mm, Dry Skin	d_3 = 5mm, Muscles

The parameters of dry skin permittivity as stated in [45] are ε_{s1} = 37.161, ε_{s2} = 70.171, ε_∞ =4.391, τ_1 = 7.42 × 10^{-12} and τ_2= 5.736 × 10^{-10}, while for the muscles the parameters as stated in [45] are ε_{s1} = 54.193, ε_{s2} = 192.873, ε_∞ =6.473, τ_1 = 6.796 × 10^{-12} and τ_2= 1.827 × 10^{-9}. The three test cases are simulated using the developed FDTD code with a two-term Debye relaxation model and the results were compared to study the effect of the dispersive material on the matching of the RFID tag antenna. The chip used in this analysis has an impedance of Z_c = (17.422−67.402j) Ω. In the FDTD simulation a cell of size $\Delta x = \Delta y = \Delta z$ =0.8 mm, 0.95 reduction factor of the CFL time step and 5,000 time steps are used. The results are shown in Fig. A.4.

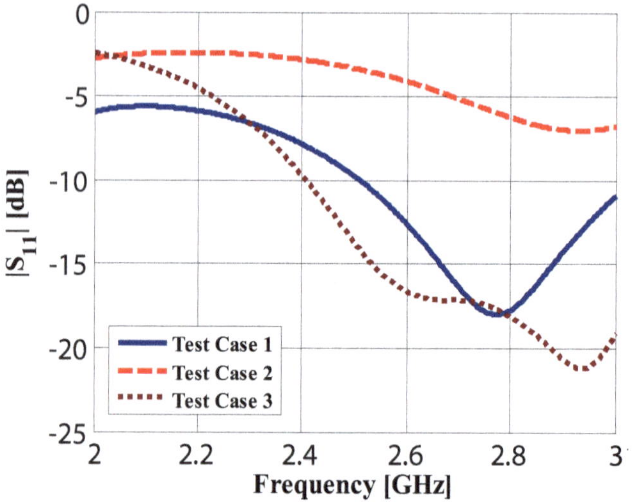

Figure A.4: Reflection coefficient magnitude of the RFID tag for the three test cases.

It should be noticed from Fig. A.4, that there exists good matching for the tag before being mounted over any human tissue, but after mounting the tag over 5 mm of dry skin the matching is degraded, which will degrade the performance of the tag significantly. However, when the tag is mounted over 5 mm of dry skin and 5 mm of muscles the matching is enhanced, which will enhance the performance of such tag. As a rule of thumb for designing a good RFID tag, the application for which this tag is going to be used on must be known. This will help the designer to know exactly the kind of substrate over which this tag will be mounted, thus an optimum tag for the application can be designed.

A.4 TRANSFORMATION FROM LORENTZ MODEL TO DEBYE MODEL FOR GOLD AND SILVER MEDIA

In this section the transformation from a single-term Lorentz model to a two-term Debye model is derived. The single-term Lorentz model can be stated as

$$\varepsilon_{rL}(\omega) = \varepsilon_\infty + \frac{(\varepsilon_s - \varepsilon_\infty)\omega_0^2}{\omega_0^2 + 2j\omega\delta_0 - \omega^2}, \tag{A.14}$$

which can be reduced to:

$$\varepsilon_{rL}(\omega) = \varepsilon_\infty + \frac{(\varepsilon_s - \varepsilon_\infty)}{1 + j\omega\left(\frac{2\delta_0}{\omega_0^2}\right) - \left(\frac{\omega^2}{\omega_0^2}\right)}, \tag{A.15}$$

$$\varepsilon_{rL}(\omega) = \frac{\varepsilon_s + j\omega\left(\frac{2\delta_0}{\omega_0^2}\varepsilon_\infty\right) - \omega^2\left(\frac{\varepsilon_\infty}{\omega_0^2}\right)}{1 + j\omega\left(\frac{2\delta_0}{\omega_0^2}\right) - \omega^2\left(\frac{1}{\omega_0^2}\right)}. \tag{A.16}$$

The two-term Debye model can be stated as

$$\varepsilon_{rD}(\omega) = \frac{\varepsilon_s + j\omega(\varepsilon_{s1}\tau_2 + \varepsilon_{s2}\tau_1) - \omega^2(\tau_1\tau_2\varepsilon_\infty)}{1 + j\omega(\tau_1 + \tau_2) - \omega^2(\tau_1\tau_2)}. \tag{A.17}$$

From (A.16) and (A.17), the following equations can be obtained:

$$\tau_1 + \tau_2 = \frac{2\delta_0}{\omega_0^2}, \quad \tau_1\tau_2 = \frac{1}{\omega_0^2}. \tag{A.18}$$

Solving these two equations simultaneously, the following relation will be obtained:

$$\tau_1 = \frac{2\delta_0 - 1}{\omega_0^2}, \quad \tau_2 = \frac{1}{2\delta_0 - 1}. \tag{A.19}$$

In addition, from (A.16) and (A.17), the following equations can be obtained:

$$\varepsilon_{s1}\tau_2 + \varepsilon_{s2}\tau_1 = \frac{2\delta_0}{\omega_0^2}\varepsilon_\infty, \tag{A.20}$$

$$\varepsilon_{s1} + \varepsilon_{s2} - \varepsilon_\infty = \varepsilon_s. \tag{A.21}$$

Solving these two equations simultaneously, one obtains:

$$\varepsilon_{s1} = \varepsilon_s + \varepsilon_\infty - \varepsilon_{s2} \tag{A.22}$$

$$\varepsilon_{s2} = \frac{\left[\frac{2\delta_0}{\omega_0^2}\varepsilon_\infty - (\varepsilon_s + \varepsilon_\infty)\,\tau_2\right]}{(\tau_1 - \tau_2)}. \tag{A.23}$$

Using Equations (A.19), (A.22), and (A.23), if the Lorentz model parameters are known the Debye model parameters can be easily calculated.

APPENDIX B

Scattering Matrix of Periodic Structures

B.1 GENERAL S- TO T-PARAMETERS TRANSFORMATION

To prove Equation (5.3) for the general case matrix, a partitioning technique should be used as follows:

$$\begin{bmatrix} b_1 \\ b_2 \\ \cdots \\ b_3 \\ b_4 \end{bmatrix} = \begin{bmatrix} S_{11} & S_{12} & \vdots & S_{13} & S_{14} \\ S_{21} & S_{22} & \vdots & S_{23} & S_{24} \\ \cdots & \cdots & \vdots & \cdots & \cdots \\ S_{31} & S_{32} & \vdots & S_{33} & S_{34} \\ S_{41} & S_{42} & \vdots & S_{43} & S_{44} \end{bmatrix} \begin{bmatrix} a_1 \\ a_2 \\ \cdots \\ a_3 \\ a_4 \end{bmatrix} \Leftrightarrow \begin{bmatrix} b_1 \\ b_2 \\ \cdots \\ a_1 \\ a_2 \end{bmatrix} = \begin{bmatrix} T_{11} & T_{12} & \vdots & T_{13} & T_{14} \\ T_{21} & T_{22} & \vdots & T_{23} & T_{24} \\ \cdots & \cdots & \vdots & \cdots & \cdots \\ T_{31} & T_{32} & \vdots & T_{33} & T_{34} \\ T_{41} & T_{42} & \vdots & T_{43} & T_{44} \end{bmatrix} \begin{bmatrix} a_3 \\ a_4 \\ \cdots \\ b_3 \\ b_4 \end{bmatrix} \tag{B.1}$$

$$|B_1> = \begin{bmatrix} b_1 \\ b_2 \end{bmatrix}, \ |B_2> = \begin{bmatrix} b_3 \\ b_4 \end{bmatrix}, \ |A_1> = \begin{bmatrix} a_1 \\ a_2 \end{bmatrix}, \ |A_2> = \begin{bmatrix} a_3 \\ a_4 \end{bmatrix}$$

$$\underset{\sim 11}{S} = \begin{bmatrix} S_{11} & S_{12} \\ S_{21} & S_{22} \end{bmatrix}, \underset{\sim 12}{S} = \begin{bmatrix} S_{13} & S_{14} \\ S_{23} & S_{24} \end{bmatrix}, \underset{\sim 21}{S} = \begin{bmatrix} S_{31} & S_{32} \\ S_{41} & S_{42} \end{bmatrix}, \underset{\sim 22}{S} = \begin{bmatrix} S_{33} & S_{34} \\ S_{43} & S_{44} \end{bmatrix}, \tag{B.2}$$

$$\underset{\sim 11}{T} = \begin{bmatrix} T_{11} & T_{12} \\ T_{21} & T_{22} \end{bmatrix}, \underset{\sim 12}{T} = \begin{bmatrix} T_{13} & T_{14} \\ T_{23} & T_{24} \end{bmatrix}, \underset{\sim 21}{T} = \begin{bmatrix} T_{31} & T_{32} \\ T_{41} & T_{42} \end{bmatrix}, \underset{\sim 22}{T} = \begin{bmatrix} T_{33} & T_{34} \\ T_{43} & T_{44} \end{bmatrix}.$$

$$\begin{bmatrix} |B_1> \\ |B_2> \end{bmatrix} = \begin{bmatrix} \underset{\sim 11}{S} & \underset{\sim 12}{S} \\ \underset{\sim 21}{S} & \underset{\sim 22}{S} \end{bmatrix} \begin{bmatrix} |A_1> \\ |A_2> \end{bmatrix},$$

$$\begin{bmatrix} |B_1> \\ |A_1> \end{bmatrix} = \begin{bmatrix} \underset{\sim 11}{T} & \underset{\sim 12}{T} \\ \underset{\sim 21}{T} & \underset{\sim 22}{T} \end{bmatrix} \begin{bmatrix} |A_2> \\ |B_2> \end{bmatrix}. \tag{B.3}$$

From Equation (B.3), four equations can be stated as follows:

$$|B_1> = \underset{\sim 11}{S}|A_1> + \underset{\sim 12}{S}|A_2> \tag{B.4a}$$

$$|B_2> = \underset{\sim 21}{S}|A_1> + \underset{\sim 22}{S}|A_2> \tag{B.4b}$$

$$|B_1> = \underset{\sim 11}{T}|A_2> + \underset{\sim 12}{T}|B_2> \tag{B.4c}$$

$$|A_1> = \underset{\sim 21}{T}|A_2> + \underset{\sim 22}{T}|B_2> \tag{B.4d}$$

To convert from S- to T-parameters we multiply Equation (B.4b) by $\underset{\sim 21}{S}^{-1}$ from the left-hand side, and then the equation will reduce to

$$\begin{aligned} &\underset{\sim 21}{S}^{-1}|B_2> = |A_1> + \underset{\sim 21}{S}^{-1}\underset{\sim 22}{S}|A_2>, \\ &|A_1> = -\underset{\sim 21}{S}^{-1}\underset{\sim 22}{S}|A_2> + \underset{\sim 21}{S}^{-1}|B_2>. \end{aligned} \tag{B.5}$$

From this equation, two T-parameters can be calculated as

$$\underset{\sim 21}{T} = -\underset{\sim 21}{S}^{-1}\underset{\sim 22}{S}, \underset{\sim 22}{T} = \underset{\sim 21}{S}^{-1}. \tag{B.6}$$

By substituting (B.5) in (B.4a) the following equation will be obtained

$$|B_1> = (\underset{\sim 12}{S} - \underset{\sim 11}{S}\underset{\sim 21}{S}^{-1}\underset{\sim 22}{S})|A_2> + \underset{\sim 11}{S}\underset{\sim 21}{S}^{-1}|B_2>. \tag{B.7}$$

From (B.7), the other two T-parameters can be calculated as

$$\underset{\sim 11}{T} = \underset{\sim 12}{S} - \underset{\sim 11}{S}\underset{\sim 21}{S}^{-1}\underset{\sim 22}{S}, \underset{\sim 12}{T} = \underset{\sim 11}{S}\underset{\sim 21}{S}^{-1}. \tag{B.8}$$

To convert from T- to S-parameters we multiply Equation (B.4d) by $\underset{\sim 22}{T}^{-1}$ from the left-hand side, and then the equation will reduce to

$$\begin{aligned} &\underset{\sim 22}{T}^{-1}|A_1> = \underset{\sim 22}{T}^{-1}\underset{\sim 21}{T}|A_2> + |B_2> \\ &|B_2> = \underset{\sim 22}{T}^{-1}|A_1> - \underset{\sim 22}{T}^{-1}\underset{\sim 21}{T}|A_2> \end{aligned} \tag{B.9}$$

From this equation, two S-parameters can be calculated as

$$\underset{\sim 21}{S} = \underset{\sim 22}{T}^{-1}, \underset{\sim 22}{S} = -\underset{\sim 22}{T}^{-1}\underset{\sim 21}{T} \tag{B.10}$$

By substituting (B.9) in (B.4c) the following equation will be obtained

$$|B_1> = \underset{\sim 12}{T}\underset{\sim 22}{T}^{-1}|A_1> + (\underset{\sim 11}{T} - \underset{\sim 12}{T}\underset{\sim 22}{T}^{-1}\underset{\sim 21}{T})|A_2>. \tag{B.11}$$

From (B.11), the other two S- parameters can be calculated as

$$\underset{\sim 11}{S} = \underset{\sim 12}{T}\underset{\sim 22}{T}^{-1}, \underset{\sim 12}{S} = (\underset{\sim 11}{T} - \underset{\sim 12}{T}\underset{\sim 22}{T}^{-1}\underset{\sim 21}{T}). \tag{B.12}$$

The general result is then given by

$$[T] = \begin{bmatrix} S_{12} - S_{11}S_{21}^{-1}S_{22} & S_{11}S_{21}^{-1} \\ -S_{21}^{-1}S_{22} & S_{21}^{-1} \end{bmatrix}, \tag{B.13a}$$

$$[S] = \begin{bmatrix} T_{12}T_{22}^{-1} & T_{11} - T_{12}T_{22}^{-1}T_{21} \\ T_{22}^{-1} & -T_{22}^{-1}T_{21} \end{bmatrix}. \tag{B.13b}$$

B.2 SQUARE PATCH MULTILAYERED FSS

To study the effect of the geometry on the harmonic frequency behavior, a test case with two identical square patch FSS layers is studied. In this test case the multilayer geometry consists of two identical FSS structures consisting of square patch elements (1:1 case). The square patch has a side length of 10 mm. The periodicity is 15 mm in both the x- and y-directions. The substrate has a thickness of 6 mm and relative permittivity ε_r = 2.2, as shown in Fig. B.1. The structure is illuminated by a normally incident plane wave and the frequency range of interest is 0-16 GHz.

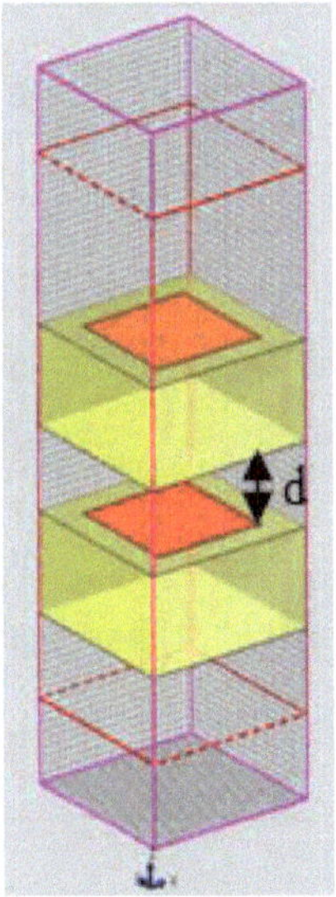

Figure B.1: The geometry of two identical square patch FSS structures.

The goal is to determine the distance d after which all the higher harmonics magnitudes reach −40dB from the magnitude of the corresponding incident field components. Using the gap determination procedure:

1) The two layers are identical, so analyzing the harmonics of one layer is enough. The reflection and transmission harmonics must be calculated.

2) The frequency range of interest as specified by the problem is 0-16G Hz (as shown in Fig. 5.9 the highest frequency will have the strongest effect).

3) k_x^i and k_y^i are equal to zero (normal incidence).

4) Determine the cut-off frequencies for the first eight harmonics as follows:

$$M_{01}, M_{0-1} \rightarrow k = \sqrt{(0)^2 + (\pm 418.9)^2} = 418.9 \rightarrow f^{01}_{turn-on} = f^{0-1}_{turn-on} = 20\text{GHz},$$
$$M_{10}, M_{-10} \rightarrow k = \sqrt{(\pm 418.9)^2 + (0)^2} = 418.9 \rightarrow f^{10}_{turn-on} = f^{-10}_{turn-on} = 20\text{GHz},$$
$$M_{11}, M_{1-1} \rightarrow k = \sqrt{(418.9)^2 + (\pm 418.9)^2} = 592.4 \rightarrow f^{11}_{turn-on} = f^{1-1}_{turn-on} = 28.3\text{GHz},$$
$$M_{-11}, M_{-1-1} \rightarrow k = \sqrt{(-418.9)^2 + (\pm 418.9)^2} = 592.4 \rightarrow f^{-11}_{turn-on} = f^{-1-1}_{turn-on} = 28.3\text{GHz}.$$

5) Use the harmonic analysis to calculate the magnitudes of the first eight harmonics, and plot the behavior of these harmonics with frequency as shown in Figs. B.2 and B.3.

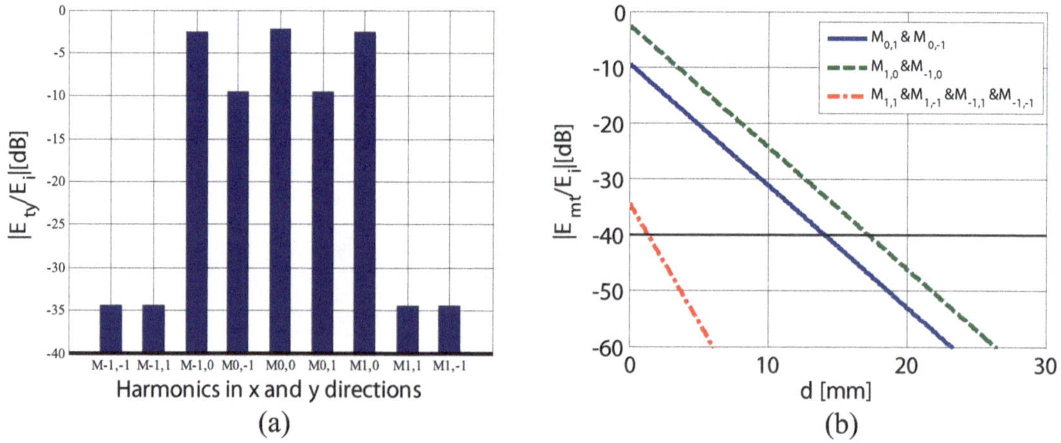

Figure B.2: The first eight transmitted harmonics at 16 GHz, (a) Magnitude normalized to incident electric field, (b) The decaying behavior with distance.

It should be noticed from Figs. B.2 and B.3 that a distance d = 17.26 mm between the two layers for this range of frequencies is enough to neglect all the harmonics. In addition, it should be noticed that the effect of the harmonics is stronger for the square-patch case compared to the dipole-FSS in Section 5.5.2, Figs. 5.16 and 5.17, which indicates the effect of the geometry on the harmonic behavior for the same periodicity (so both the geometry and periodicity controls the harmonics behavior).

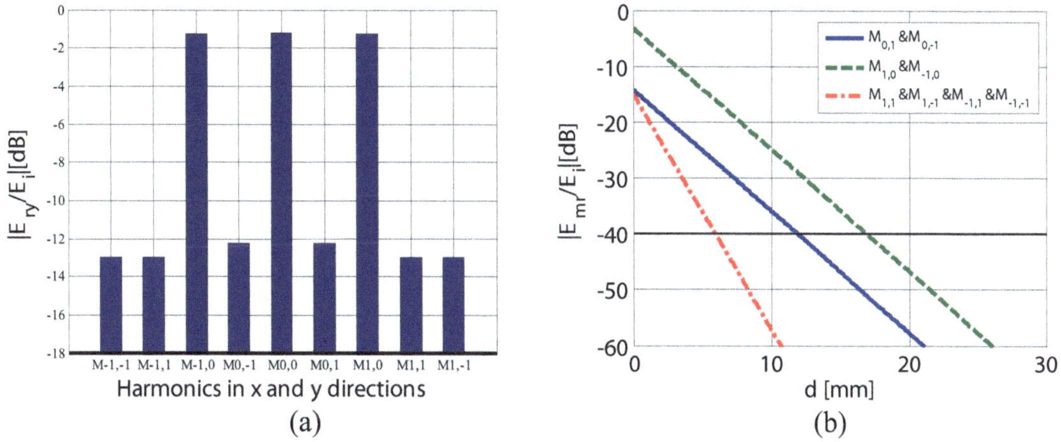

Figure B.3: The first eight reflected harmonics at 16 GHz, (a) Magnitude normalized to incident electric field, (b) The decaying behavior with distance.

B.3 L-SHAPED MULTILAYERED FSS

To study the effect of cross-polarized fields, a test case with two identical L-shaped FSS structures is studied (1:1 case). The L-shaped element consists of two perpendicular dipoles of length 12 mm and width 3 mm. The periodicity is 15 mm in both x- and y-directions. The substrate has thickness of 6 mm and relative permittivity ε_r = 2.2, as shown in Fig. B.4. The structure is illuminated by

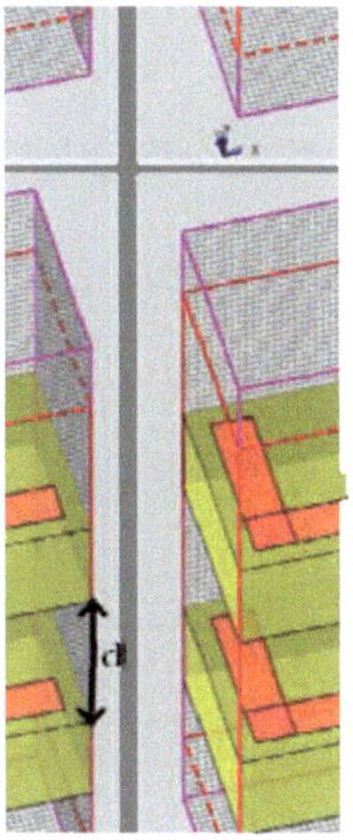

Figure B.4: Two identical L-shaped FSS geometry.

a normally incident plane wave and the frequency range of interest is 0-15 GHz. This structure is used to study the effect of cross-polarization components. The structure generates cross-polarization components with normal incidence and the reflection and transmission co- and cross-polarization coefficients are shown in Fig. B.5. Using the harmonic analysis, the proper distance for ignoring the

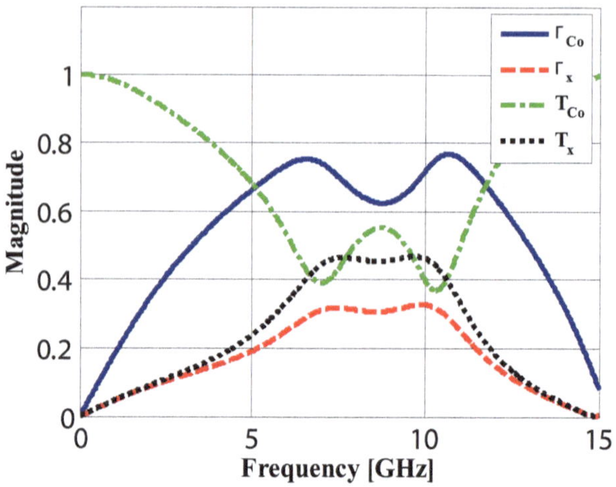

Figure B.5: Reflection and transmission coefficients for co- and cross-polarized components for single layer L-shaped FSS structure.

harmonics was found to be 13.34 mm. The structure is simulated using the cascaded technique with $d = 15$ mm. The results are compared to the FDTD simulation of the entire structure, as shown is Fig. B.6. A good agreement is observed between the results obtained from two approaches.

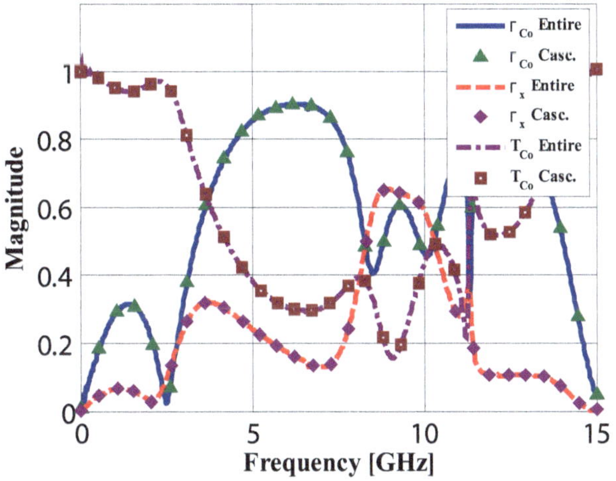

Figure B.6: Reflection and transmission coefficients for co- and cross-polarized components for two identical L-shaped FSS structures with $d = 15$ mm.

References

[1] A. Z. Elsherbeni and V. Demir, *The Finite-Difference Time-Domain Method for Electromagnetics: With MATLAB Simulations*. SciTech Publishing, 2009. Cited on page(s) 1, 6, 7, 11, 19, 95, 96

[2] J. Maloney, and M. P. Kesler, *Analysis of Periodic Structures, Computational Electrodynamics: The Finite-Difference Time-Domain Method,* 3rd ed. Norwood, MA: Artech House, 2005. Cited on page(s) 1

[3] J. A. Roden, S. D. Gedney, P. Kesler, J. G. Maloney, and P. H. Harms, "Time-domain Analysis of Periodic Structures at Oblique Incidence: Orthogonal and Nonorthogonal FDTD Implementations," *IEEE Transaction on Antennas and Propagation*, vol. 46, DOI: 10.1109/22.664143 Cited on page(s) 1

[4] Y. C. A. Kao and R. G. Atkins, "A Finite-Difference Time-Domain Approach for Frequency Selective Surfaces at Oblique Incidence," *IEEE Antennas and Propagation Society International Symposium*, vol. 2, pp. 21–26, 1996. DOI: 10.1109/APS.1996.549866 Cited on page(s) 1

[5] P. Harms, R. Mittra, and W. Ko, "Implementation of The Periodic Boundary Condition in the Finite-Difference Time-Domain Algorithm for FSS Structures," *IEEE Transaction on Antennas and Propagation*, vol. 42, no. 9, pp. 1317–1324, 1994. DOI: 10.1109/8.318653 Cited on page(s) 1

[6] F. Yang, J. Chen, R. Qiang, and A. Z. Elsherbeni, "A Simple and Efficient FDTD/PBC Algorithm for Scattering Analysis of Periodic Structures," *Radio Science*, vol. 42, RS4004, 2007. DOI: 10.1029/2006RS003526 Cited on page(s) 2, 13, 14, 67

[7] F. Yang, J. Chen, R. Qiang, and A. Z. Elsherbeni, "FDTD Analysis of Periodic Structures at Arbitrary Incidence Angles: A Simple and Efficient Implementation of the Periodic Boundary Conditions," *IEEE Antennas and Propagation Society International Symposium,* pp. 2715–2718, 2006. DOI: 10.1109/APS.2006.1711164 Cited on page(s) 2

[8] F. Yang, A. Z. Elsherbeni, and J. Chen, "A Hybrid Spectral-FDTD/ARMA Method for Periodic Structure Analysis," *IEEE Antennas and Propagation Society International Symposium*, pp. 3720–3723, 2007. DOI: 10.1109/APS.2007.4396347 Cited on page(s) 2

[9] S. Xiao, R. Vahldieck, and H. Jin, "Full-wave Analysis of Guided Wave Structures using A Novel 2-D FDTD," *IEEE Microwave Guided Wave Letters*, 2(5), 165–167, 1992. DOI: 10.1109/75.134342 Cited on page(s) 2

[10] A. Aminian, and Y. Rahmat-Samii, "Spectral FDTD: A Novel Computational Technique for the Analysis of Periodic Structures," *IEEE Antennas and Propagation Society International Symposium*, no. 3, pp. 3139–3142, 2004. DOI: 10.1109/APS.2004.1332044 Cited on page(s) 2

[11] A. Aminian, F. Yang, and Y. Rahmat-Samii, "Bandwidth Determination for Soft and Hard Ground Planes by Spectral FDTD: A Unified Approach in Visible and Surface Wave Regions," *IEEE Transaction on Antennas and Propagation*, vol. 53, no. 1, pp. 18–28, 2005. DOI: 10.1109/TAP.2004.840517 Cited on page(s) 2

[12] A. Aminian, and Y. Rahmat-Samii, "Spectral FDTD: A Novel Technique for the Analysis of Oblique Incident Plane Wave on Periodic Structures," *IEEE Transaction on Antennas and Propagation*, vol. 54, no. 6, pp. 1818–1825, 2006. DOI: 10.1109/TAP.2006.875484 Cited on page(s) 2

[13] K. S. Yee, "Numerical Solution of Initial Boundary Value Problems Involving Maxwell's Equations in Isotropic Media," *IEEE Transaction on Antennas and Propagation*, vol. AP-14, pp. 302–307, 1966. DOI: 10.1109/TAP.1966.1138693 Cited on page(s) 6

[14] G. Mur, "Absorbing Boundary Conditions for the Finite-Difference Approximation of The Time-Domain Electromagnetic-Field Equations," *IEEE Transactions on Electromagnetic Compatibility*, vol. EMC-23, pp. 377–382, 1981. DOI: 10.1109/TEMC.1981.303970 Cited on page(s) 10

[15] Z. P. Liao, H. L. Wong, B. Yang, and Y. Yuan, "A Transmitting Boundary for Transient Wave Analysis," *Sci. Sin., Ser. A*, vol. 27, pp. 1063–1076, 1984. Cited on page(s) 10

[16] J. P. Berenger, "A Perfectly Matched Layer for The Absorption of Electromagnetic Waves," *Journal of Computational Physics*, vol. 114, pp. 185–200, 1994. DOI: 10.1006/jcph.1994.1159 Cited on page(s) 10

[17] J. Roden and S. Gedney, "Convolutional PML (CPML): An Efficient FDTD Implementation of The CFS-PML for Arbitrary Media," *Microwave and Optical Technology Letters*, vol. 27, no. 5, pp. 334–339, 2000.
DOI: 10.1002/1098-2760(20001205)27:5%3C334::AID-MOP14%3E3.3.CO;2-1 Cited on page(s) 10

[18] A. Taflove and S. C. Hagness, *Computational Electrodynamics: The Finite-Difference Time-Domain Method.* Chapter 13, 3rd ed. Norwood, MA: Artech House, 2005. Cited on page(s) 13, 69

[19] MATLAB software is distributed by mathworks `http://www.mathworks.com/` Cited on page(s) 19, 35

[20] Ansoft Designer Software is distributed by the Ansoft Corporation: http://www.ansoft.com/products/hf/ansoft_designer/ Cited on page(s) 19

[21] R. Courant, K. Friedrichs, and H. Lewy, "On the Partial Difference Equations of Mathematical Physics," *IBM Journal of Research and Development*, vol. 11, no. 2, pp. 215–234, 1967. DOI: 10.1147/rd.112.0215 Cited on page(s) 19

[22] B. A. Munk, *Frequency Selective Surface*. John Wiley, New York, 2000. DOI: 10.1002/0471723770 Cited on page(s) 21, 27

[23] I. Stevanovic', P. Crespo-Valero, K. Blagovic', F. Bongard, and J. R. Mosig, "Integral-Equation Analysis of 3-D Metallic Objects A rranged in 2-D Lattices Using the Ewald Transformation," *IEEE Transaction on Microwave Theory and Techniques*, vol. 54, no. 10, pp. 3688–3697, 2006. DOI: 10.1109/TMTT.2006.882876 Cited on page(s) 24

[24] Z. Yun and M. F. Iskander, "Implementation of Floquet Boundary Conditions in FDTD Analysis of Periodic Phased Array Antennas with Skewed Grid," *Electromagnetics Taylor and Francis*, vol. 20, Issue 5, pp. 445–452, 2000. DOI: 10.1080/027263400750064437 Cited on page(s) 27

[25] K. ElMahgoub, F. Yang, A. Z. Elsherbeni, V. Demir, and J. Chen "FDTD Analysis of Periodic Structures with Arbitrary Skewed Grid," *IEEE Transaction on Antennas and Propagation*, vol. 58, no. 8, pp. 2649–2657, 2010. DOI: 10.1109/TAP.2010.2050445 Cited on page(s) 27, 28, 29, 32, 36, 37, 38

[26] K. ElMahgoub, F. Yang, A. Z. Elsherbeni, V. Demir, and J. Chen, "FDTD/PBC algorithm for skewed grid periodic structures," IEEE Antennas and Propagation Society International Symposium, 2010. DOI: 10.1109/APS.2010.5562016 Cited on page(s) 28, 39, 40

[27] R. Luebbers, F. P. Hunsberger, K. S. Kunz, R. B. Standler, and M. Schneider, "A Frequency-Dependent Finite-Difference Time-Domain Formulation for Dispersive Materials," *IEEE Transactions on Electromagnetic Compatibility*, vol. 32, no. 3, pp. 222–227, 1990. DOI: 10.1109/15.57116 Cited on page(s) 41

[28] M. D. Bui, S. S. Stuchly, and G. I. Costache, "Propagation of Transients in Dispersive Dielectric Media," *IEEE Transaction on Microwave Theory and Techniques*, vol. 39, no. 7, pp. 1165–1172, 1991. DOI: 10.1109/22.85384 Cited on page(s) 41

[29] D. M. Sullivan, "A Frequency-Dependent FDTD Method for Biological Applications," *IEEE Transaction on Microwave Theory and Techniques*, vol. 40, no. 3, pp. 532–539, 1992. DOI: 10.1109/22.121729 Cited on page(s) 41

[30] R. J. Luebbers, F. P. Hunsberger, and K. S. Kunz, "A Frequency-Dependent Finite-Difference Time-Domain Formulation for Transient Propagation in Plasma," *IEEE Transaction on Antennas and Propagation*, vol. 39, no. 1, pp. 29–34, January 1991. DOI: 10.1109/8.64431 Cited on page(s) 41

[31] R. J. Luebbers and F. P. Hunsberger, "FDTD for Nth-order Dispersive Media," *IEEE Transaction on Antennas and Propagation*, vol. 40, no. 11, pp. 1297–1301, 1992. DOI: 10.1109/8.202707 Cited on page(s) 41

[32] R. J. Luebbers, D. Steich, and K. S. Kunz, "FDTD Calculation of Scattering from Frequency-Dependent Materials," *IEEE Transaction on Antennas and Propagation*, vol. 41, no. 9, pp. 1249–1257, 1993. DOI: 10.1109/8.247751 Cited on page(s) 41

[33] T. Kashiwa and I. Fukai, "A Treatment by The FD-TD Method of The Dispersive Characteristics Associated with Electronic Polarization," *Microwave Optical Technology Letters*, vol. 3, no. 6, pp. 203–205, 1990. DOI: 10.1002/mop.4650030606 Cited on page(s) 41

[34] R. M. Joseph, S. C. Hagness, and A. Taflove, "Direct Time Integration of Maxwell's Equations in Linear Dispersive Media with Absorption for Scattering and Propagation of Femtosecond Electromagnetic Pulses," *Optics Letters*, vol. 16, no. 9, pp. 1412–1414, 1991. DOI: 10.1364/OL.16.001412 Cited on page(s) 41

[35] D. F. Kelley and R. J. Luebbers, "Piecewise Linear Recursive Convolution for Dispersive Media using FDTD," *IEEE Transaction Antennas and Propagation*, vol. 44, no. 6, pp. 792–798, 1996. DOI: 10.1109/8.509882 Cited on page(s) 41

[36] D. M. Sullivan, "Frequency-Dependent FDTD Methods Using Z transforms," *IEEE Transaction Antennas and Propagation*, vol. 40, no. 10, pp. 1223–1230, 1992. DOI: 10.1109/8.182455 Cited on page(s) 41

[37] D. M. Sullivan, "Nonlinear FDTD Formulations Using Z Transforms," *IEEE Transaction on Microwave Theory and Techniques*, vol. 43, no. 3, pp. 676–682, 1995. DOI: 10.1109/22.372115 Cited on page(s) 41

[38] D. M. Sullivan, "Z-Transform Theory and The FDTD Method," *IEEE Transaction on Antennas and Propagation*, vol. 44, no. 1, pp. 28–34, 1996. DOI: 10.1109/8.477525 Cited on page(s) 41, 99

[39] A. Taflove and S. C. Hagness, *Computational Electrodynamics: The Finite-Difference Time-Domain Method.* Chapter 9, 3rd edition, Norwood, MA: Artech House, 2005. Cited on page(s) 41

[40] O. P. Gandhi, B. Gao, and J. Chen, "A Frequency-Dependent Finite-Difference Time-Domain Formulation for General Dispersive Media," *IEEE Transaction on Microwave Theory and Techniques*, vol. 41, no. 4, pp. 658–665, 1993. DOI: 10.1109/22.231661 Cited on page(s) 41, 51

[41] C. M. Furse, J. Chen, and O. P. Gandhi, "The Use of the Frequency Dependent Finite-Difference Time-Domain Method for Induced Current and SAR Calculations for a Heterogeneous Model of the Human Body," *IEEE Transaction on Electromagnetic Compatibility*, vol. 36, no. 2, pp. 128–133, 1994. DOI: 10.1109/15.293278 Cited on page(s) 41

[42] H. H. Abdullah, F. M. El-Hefnawi, and A. Z. Elsherbeni, "A FDTD Scattered Field Formulation for Dispersive Media," IEEE Antennas and Propagation Society International Symposium, 2000. DOI: 10.1109/APS.2000.873755 Cited on page(s) 95

[43] Ansoft HFSS Software is distributed by the Ansoft Corporation: `http://www.ansoft.com/products/hf/ansoft_hfss/` Cited on page(s) 51

[44] K. ElMahgoub, A. Z. Elsherbeni, and F. Yang, "Analysis of periodic structures with dispersive material using the FDTD technique," XXXth URSI General Assembly and Scientific Symposium, 2011. DOI: 10.1109/URSIGASS.2011.6050393 Cited on page(s) 51, 52, 53, 57, 58, 59

[45] M. A. Eleiwa and A. Z. Elsherbeni. "Debye Constants for Biological Tissues from 30 Hz to 20 GHz," *ACES Journal*, vol. 16, no. 3, pp. 202–213, Nov. 2001. Cited on page(s) 101

[46] Y. Zhao and Y. Hao, "Finite-Difference Time-Domain Study of Guided Modes in Nanoplasmonic Waveguides," *IEEE Transaction on Antennas and Propagation*, vol. 55, no. 11, pp. 3070–3077, 2007. DOI: 10.1109/TAP.2007.908370 Cited on page(s) 54

[47] J. Olkkonen, "FDTD Scattered Field Formulation for Scatterers in Stratified Dispersive Media," *OPTICS EXPRESS*, vol. 18, no. 5, pp. 4380–4389, 2010. DOI: 10.1364/OE.18.004380 Cited on page(s) 54

[48] P. B. Johnson and R. W. Christy, "Optical Constants of the Noble Metals," *Phys. Rev.,B, Solid State* **6**(12), pp. 4370–4379, 1972. DOI: 10.1103/PhysRevB.6.4370 Cited on page(s) 54

[49] R. Qiang, J. Chen, J. Huang, M. Koledintseva, E. Dubroff, J. Drewniak, and F. Yang "Numerical Analysis of Sandwiched Composite-FSS Structures," pp. 742–746, *IEEE, EMC 2006.* Cited on page(s) 57

[50] R. C. Hall, R. Mittra, and K. M. Mitzner, "Analysis of Multilayered Periodic Structures using Generalized Scattering Matrix Theory," *IEEE Transaction on Antennas and Propagation*, vol. 36, no. 4, pp. 511–517, 1988. DOI: 10.1109/8.1140 Cited on page(s) 61, 69

[51] R. Pous and D. M. Pozar, "A Frequency-Selective Surface Using Aperture Coupled Microstrip Patches," *IEEE Transaction on Antennas and Propagation*, vol. 39, no. 12, pp. 1763–1769, 1991. DOI: 10.1109/8.121598 Cited on page(s) 61

[52] R. Mittra and S. W. Lee, *Analytical Techniques in the Theory of Guided Waves*, Macmillan, 1971. Cited on page(s) 61

[53] R. E. Collin, *Foundations for Microwave Engineering,* New York: McGraw-Hill, 1966. Cited on page(s) 61

[54] W. Lee, G. Zarrillo, and C. L. Law, "Simple Formulas for Transmission Through Periodic Metal Grids or Plates," *IEEE Transaction on Antennas and Propagation,* vol. 30, no. 5, pp. 904–909, Sept. 1982. DOI: 10.1109/TAP.1982.1142923 Cited on page(s) 61

[55] C. Wan and J. A. Encinar, "Efficient Computation of Generalized Scattering Matrix for Analyzing Multilayered Periodic Structures," *IEEE Transaction on Antennas and Propagation,* vol. 43, no. 11, pp. 1233–1242, Nov. 1995. DOI: 10.1109/8.475095 Cited on page(s) 61

[56] R. F. Harrington, *Time-Harmonics Electromagnetic Fields,* New York: McGraw-Hill, 1961. Cited on page(s) 61

[57] T. Shibata, and T. Itoh, "Generalized-Scattering-Matrix Modeling of Waveguide Circuits Using FDTD Field Simulations," *IEEE Transaction on Microwave Theory and Techniques,* vol. 46, no. 11, pp. 1742–1751, Nov. 1998. DOI: 10.1109/22.734574 Cited on page(s) 67

[58] T. Huang, B. Houshmand, and T. Itoh, "Efficient Modes Extraction and Numerically Exact Matched Sources for A Homogeneous Waveguide Cross-Section in A FDTD Simulation," *IEEE Microwave Theory and Techniques Symposium Digest,* 1994. DOI: 10.1109/MWSYM.1994.335368 Cited on page(s) 67

[59] K. ElMahgoub, F. Yang, and A. Z. Elsherbeni, "Efficient Analysis of Multi-Layer Periodic Structures Using FDTD," *Antenna Measurement Techniques Association Conference,* 2010. Cited on page(s) 80

[60] K. ElMahgoub, F. Yang, and A. Z. Elsherbeni, "Analysis of Multilayer Periodic Structures with Different Periodicities using a Hybrid FDTD/GSM Method," *IEEE International Microwave Symposium,* Baltimore, 2011. DOI: 10.1109/MWSYM.2011.5972914 Cited on page(s) 85, 86, 88

Authors' Biographies

KHALED ELMAHGOUB

Dr. Khaled ElMahgoub received B.Sc. and M.Sc. degrees in electronics and electrical communications engineering from Cairo University, Egypt, in 2001 and 2006, respectively. He received his Ph.D. in electrical engineering from the University of Mississippi, USA in 2010. From 2007–2010, he was a teaching and research assistant at the University of Mississippi. Prior to that, from 2001–2006 he has been a teaching and research assistant at Cairo University. Currently, he is working as senior validation engineer at Trimble Navigation, Cambridge, MA, USA, possessing over six years of experience in the industry. Throughout his academic years, he coauthored over 20 technical journals and conference papers. He is the main co-author of the book entitled *Enhancements to Low Density Parity Check Codes: Application to the WiMAX System*, Lambert Academic Publishing, 2010. ElMahgoub is editor assistant for *Applied Computational Electromagnetics Society (ACES) Journal*. He is also a frequent reviewer for many scientific journals, conferences, books, and has chaired technical sessions in international conferences. He is a member of IEEE, ACES, and Phi Kappa Phi honor society. His current research interests include RFID systems, channel coding, FDTD, antenna design, and numerical techniques for electromagnetics.

FAN YANG

Prof. Fan Yang received B.S. and M.S. degrees from Tsinghua University, Beijing, China, in 1997 and 1999, respectively, and a Ph.D. from the University of California at Los Angeles (UCLA) in 2002.

From 1994–1999, he was a Research Assistant at the State Key Laboratory of Microwave and Digital Communications, Tsinghua University. From 1999–2002, he was a Graduate Student Researcher at the Antenna Laboratory, UCLA. From 2002–2004, he was a Post-Doctoral Research Engineer and Instructor at the Electrical Engineering Department, UCLA. In August 2004, he joined the Electrical Engineering Department, University of Mississippi, as an Assistant Professor, and was promoted to an Associate Professor. In 2010, he became a Professor at the Electronic Engineering Department, Tsinghua University.

Prof. Yang's research interests include antenna theory, designs and measurements, electromagnetic bandgap (EBG) structures and their applications, computational electromagnetics and optimization techniques, and applied electromagnetic systems such as the radio frequency identification (RFID) system and solar energy system. He has published over 150 journal and conference papers, 5 book chapters, and 2 books entitled *Electromagnetic Band Gap Structures in Antenna Engineering* (Cambridge University Press, 2009) and *Electromagnetics and Antenna Optimization Using Taguchi's Method* (Morgan & Claypool, 2007).

Dr. Yang is a Senior Member of IEEE, and was secretary of the IEEE Antennas and Propagation Society, Los Angeles Chapter. He is a member of URSI-USNC. He serves as an Associate Editor for the *IEEE Transactions on Antennas And Propagation* and *Applied Computational Electromagnetics Society (ACES) Journal*. He is also a frequent reviewer for over 20 scientific journals and book publishers, and has chaired technical sessions in numerous international symposia. Prof. Yang has been the recipient of several prestigious awards and recognitions, including the 2004 Certificate for Exceptional Accomplishment in Research and Professional Development of UCLA, the Young Scientist Award of the 2005 URSI General Assembly and of the 2007 International Symposium on Electromagnetic Theory, the 2008 Junior Faculty Research Award of the University of Mississippi, and the 2009 inaugural IEEE Donald G. Dudley Jr. Undergraduate Teaching Award.

ATEF Z. ELSHERBENI

IEEE Fellow (2007) and ACES Fellow (2008)

Finland Distinguished Professor (2009)

Dr. Atef Z. Elsherbeni received an honor B.Sc. degree in Electronics and Communications, an honor B.Sc. degree in Applied Physics, and a M.Eng. degree in Electrical Engineering, all from Cairo University, Cairo, Egypt, in 1976, 1979, and 1982, respectively. Ph.D. degree in Electrical Engineering from Manitoba University, Winnipeg, Manitoba, Canada in 1987. He was a part-time Software and System Design Engineer from March 1980–December 1982 at the Automated Data System Center, Cairo, Egypt. From January–August 1987, he was a Post Doctoral Fellow at Manitoba University. Dr. Elsherbeni joined the faculty at the University of Mississippi in August 1987 as an Assistant Professor of Electrical Engineering. He advanced to the rank of Associate Professor in July 1991, and to the rank of Professor in July 1997. He became the director of The School of Engineering CAD Lab on August 2002, and the director of the Center for Applied Electromagnetic Systems Research (CAESR) in July 2011. In July 2009 he was appointed Associate Dean of Engineering for Research and Graduate Programs at the University of Mississippi. He was appointed Adjunct Professor at The Department of Electrical Engineering and Computer Science of the L.C. Smith College of Engineering and Computer Science at Syracuse University in January 2004. He spent a sabbatical term in 1996 at the Electrical Engineering Department, University of California at Los Angeles (UCLA) and was a visiting Professor at Magdeburg University during the summer of 2005 and at Tampere University of Technology in Finland during the summer of 2007. In 2009 he was selected as Finland Distinguished Professor by the Academy of Finland and TEKES.

Dr. Elsherbeni received the 2006 and 2011 School of Engineering Senior Faculty Research Award for Outstanding Performance in research, the 2005 School of Engineering Faculty Service Award for Outstanding Performance in Service, the 2004 Valued Service Award from the Applied Computational Electromagnetics Society (ACES) for Outstanding Service as 2003 ACES Symposium Chair, the Mississippi Academy of Science 2003 Outstanding Contribution to Science Award, the 2002 IEEE Region 3 Outstanding Engineering Educator Award, the 2002 School of Engineering Outstanding Engineering Faculty Member of the Year Award, the 2001 ACES Exemplary Service Award for leadership and contributions as Electronic Publishing Managing Editor 1999–2001, the 2001 Researcher/Scholar of the year award in the Department of Electrical Engineering, The University of Mississippi, and the 1996 Outstanding Engineering Educator of the IEEE Memphis Section.

Over the last 24 years, Dr. Elsherbeni participated in acquiring over 10 million dollars to support his research dealing with scattering and diffraction by dielectric and metal objects, finite difference time domain analysis of passive and active microwave devices including planar transmission lines, field visualization and software development for EM education, interactions of electromagnetic waves with human body, sensors development for monitoring soil moisture, airport noise levels, air quality including haze and humidity, reflector and printed antennas and antenna arrays for radars, UAV, personal communication systems, antennas for wideband applications, and antenna and material properties measurements. He has co-authored 138 technical journal articles, 25 book chapters, and contributed to over 330 professional presentations, and offered 26 short courses and 29 invited seminars. He is the coauthor of the books entitled *The Finite Difference Time Domain Method for Electromagnetics with Matlab Simulations* (Scitech, 2009), *Antenna Design and Visualization Using Matlab* (Scitech, 2006), *MATLAB Simulations for Radar Systems Design* (CRC Press, 2003), *Electromagnetic Scattering Using the Iterative Multiregion Technique* (Morgan & Claypool, 2007), and *Electromagnetics and Antenna Optimization using Taguchi's Method* (Morgan & Claypool, 2007), and the primary author of the chapters "Handheld Antennas" and "The Finite Difference Time Domain Technique for Microstrip Antennas" in *Handbook of Antennas in Wireless Communications* (CRC Press, 2001), he was the advisor/co-advisor for 33 M.S. and 13 Ph.D. students.

Dr. Elsherbeni is a Fellow member of the Institute of Electrical and Electronics Engineers (IEEE) and a fellow member of The Applied Computational Electromagnetic Society (ACES). He is the Editor-in-Chief for *ACES Journal*, and a past Associate Editor of the *Radio Science Journal*. He serves on the editorial board of the Book Series on Progress in Electromagnetic Research, and the *Electromagnetic Waves and Applications Journal*. He was the Chair of the Engineering and Physics Division of the Mississippi Academy of Science and was the Chair of the Educational Activity Committee for the IEEE Region 3 Section.

Dr. Elsherbeni's homepage can be found at `http://www.engineering.olemiss.edu/~atef/` and his email address is `Elsherbeni@ieee.org`